A PRELIMINARY DISCOURSE

ON THE STUDY OF

NATURAL PHILOSOPHY

A PRELIMINARY DISCOURSE ON THE STUDY OF NATURAL PHILOSOPHY

John F. W. Herschel

With a new Foreword by
Arthur Fine

The University of Chicago Press

Chicago & London

This is a facsimile of the 1830 edition published as volume I of Dionysius Lardner's *Cabinet Cyclopaedia* (1832).

The University of Chicago Press, Chicago 60637
The University of Chicago Press, Ltd., London

First published in 1830
University of Chicago Press Edition 1987
Printed in the United States of America

95 94 93 92 91 90 89 88 87 5 4 3 2 1

Library of Congress Cataloging in Publication Data

Herschel, John F. W. (John Frederick William), Sir, 1792–1871.
A preliminary discourse on the study of natural philosophy.

Reprint. Originally published: London : Printed for Longman, Rees, Orme, Brown, Green, and J. Taylor, 1830. (The Cabinet cyclopaedia ; v. 1)
Includes index.
1. Science—Philosophy. 2. Science—History. 3. Physics—History. 4. Astronomy—History. I. Title. II. Series: Cabinet cyclopaedia ; v. 1.
Q175.H427 1987 501 87–5985
ISBN 0–226–32777–9 (pbk.)

FOREWORD

In some eras the popular image of science is associated with a single name: Newton in the seventeenth century, Einstein in our own. Nineteenth-century England is represented by John Herschel. A brilliant mathematician, Herschel was also an accomplished chemist, whose investigations at the interface of chemistry and optics paved the way for the development of modern photography (a word coined by Herschel in 1839). His most acclaimed successes, however, were in astronomy, especially for studies of double stars, where his work was recognized in prizes awarded by the French Academy, the Royal Society, the Astronomical Society, and others. In these studies Herschel developed and applied new graphical methods for showing how the relative motion of the two stars in a binary system closely followed the laws of Newtonian celestial mechanics. In Herschel's time this accomplishment was widely regarded as the final triumph in the demonstration that the laws of physics had universal scope.

Herschel was a person of great learning and energy, and he complemented his scientific research with an active engagement in issues of educational reform and public policy (being an early advocate of decimal coinage). His writings include hundreds of scientific papers, seven books, and several important encyclopedia articles. He also translated a number of literary works, among them some of Schiller's poems put into English hexameter, Dante's *Inferno*, and

Homer's *Iliad.* Two of the encyclopedia articles were commissioned as separate volumes in Dionysius Lardner's *Cabinet Cyclopedia,* a work intended for a semi-popular audience including the serious "amateurs" of nineteenth-century science. One of these *Cyclopedia* volumes developed into a celebrated treatise, Herschel's *Outlines of Astronomy* (1849). The other, volume one of the *Cyclopedia,* was reprinted several times, widely read by his peers, and became a standard reference work on the methodology of science. It is Herschel's *A Preliminary Discourse on the Study of Natural Philosophy* (1830).

John Herschel (Sir John Frederick William, 1792–1871) was born in Slough, England, in the county of Buckinghamshire on March 7, 1792. He was an only child, the son of Sir William Herschel (1738–1822) and the former Mary Baldwin Pitt. His father, Sir William, was the discoverer of the planet Uranus and one of the most illustrious astronomers of his time. John Herschel married relatively late in life, in 1829, to Margaret Brodie Stewart, with whom he had twelve children. Perhaps following Newton's example, Herschel's last important public activity was as master of the mint (1850–56). After holding that office he retired to work on catalogues of nebulae, star clusters, and double stars. He died on May 11, 1871, was mourned by the nation, and was buried next to Newton in Westminster Abbey.

For the most part John Herschel's early education was conducted privately near his home, although he did attend Eton, briefly, around age eight. While at home he was certainly influenced by the astronomical industry carried on there by his father, Sir William, and his aunt, Caroline Lucretia Herschel (1750–1848). At age seventeen Herschel entered St. John's College, Cambridge, where he concentrated on the study of mathematics. He completed his examinations at Cambridge in 1813, taking honors as "first senior wrangler," the highest distinction awarded. In the same year St. John's elected him a fellow, a position he retained until his marriage. Also in 1813 he was elected a fellow of the Royal Society for papers in applied mathematics, the last of which his father had the pleasure of presenting to the society.

During Herschel's time at Cambridge, he made the acquaintance of William Whewell (1794–1866), also, then, a student of mathematics and later, more philosopher than scientist, the author of a competing popular treatise on scientific method, *The History of the Inductive Sciences* (1857). Although they disagreed on the relative weight to be assigned the empirical versus the ideational elements in science, with Herschel more on the Humean, empirical end of the spectrum and Whewell more on the Kantian side, these two men were to become, in Whewell's words, "friends of a lifetime." While at Cambridge Herschel also formed a lasting friendship with Charles Babbage (1792–1871), whom we recognize today as an important modern pioneer in the design of calculating machines, and with George Peacock (1791–1858), an educational reformer and later the Lowndean Professor of Astronomy at Cambridge. These three founded the Analytical Society devoted, as Herschel would say, to the replacement of "dot-age" by "d-ism"; that is, to replacing the fluxional dot notation ($\dot{y}$) of Newton, in the differential calculus, by the (dy/dx) notation of Leibniz. Thus Herschel and his friends led the reform that opened Great Britain to the important discoveries and methods of continental analysis which, in turn, gave rise to the brilliant nineteenth-century school of British mathematical physics (A. Cayley, G. Green, W. Hamilton, J.C. Maxwell, G.G. Stokes, J.T. Sylvester, W. Thompson, and others).

In 1814 Herschel followed the example of several Cambridge mathematicians who had pursued successful careers in the law, and moved to London to begin a legal apprenticeship at Lincoln's Inn. In London, however, he became acquainted with and influenced by William Hyde Wollaston, whose scientific work, like Herschel's, included important discoveries in chemistry (e.g., the elements palladium and rhodium) and in optics (e.g., the dark lines in the solar spectrum, and the invention of the reflecting goniometer and of the camera lucida). There he also met the wealthy amateur astronomer James South, with whom he later (1821–23) collaborated in reexamining and extending some of William Herschel's observations on double stars, work that was recognized by prizes both from the French Academy (1825) and from the Astronomical Society (1826).

In 1815 John Herschel returned from London to Cambridge, where he applied for the chair in chemistry. Failing in that, he took up a teaching post as sublector at St. John's, pursuing his researches there and taking an M.A. in 1816. Herschel then returned home to Slough. As he later said, he was most inclined toward the study of chemistry and optics but increasingly he found the influence of his father, and the call of filial duty, moving him ever more strongly toward astronomy. Indeed in 1820, with his father's help, Herschel completed a mirror for a reflecting telescope, eighteen inches in diameter and with a focal length of twenty feet. Subsequently refined, it was with this instrument that Herschel made his most important astronomical observations.

From 1816 to 1833 Herschel carried on research in mathematics, optics, chemistry, and astronomy. In 1821 the Royal Society presented him with its Copley medal for his mathematical contributions. His optical investigations included a searching theoretical analysis of the wave versus the particle theory of light, presented in articles on light (1827) and on sound (1830) for the *Encyclopedia Metropolitana.* He also investigated properties of compound lenses, the polarization and birefringence of crystals, and the interference of light and sound. In 1819 Herschel discovered that the salts of silver were soluble in "hyposulphite of soda" (the older name for $Na_2S_2O_3.5H_2O$, sodium thiosulphate penhydrate). This is the photographer's "hypo," whose use Herschel pioneered. (He is also responsible for the photographic terminology of "positive" and "negative.") During this period Herschel continued with his astronomical observations at Slough, following up his father's work on the observation of nebulae, clusters, and double stars. His memoir *On the investigation of the orbits of double stars* won a medal from the Royal Society in 1833, and also much popular acclaim. That same year Herschel brought out a catalogue of 2,307 nebulae and clusters, and by 1836 he had issued no less than six catalogues of double stars (3,346 systems, in all). Somehow, amid all of this scientific industry Herschel found time for travel abroad, where he mixed walking tours and sightseeing with visits to virtually all the important European scientists of his generation. He also found time for marriage and family life.

In 1833 Herschel resolved to complete a survey of the heavens by doing for the southern skies what he and his father had achieved for the north. So he proceeded to move himself, his own telescope ("so as to give a unity to the results of both portions of the survey") and his family to Africa, settling in the colony near Cape Town. For the next five years he carried on his astronomical work there, along with activities in botany and meteorology. He also helped devise a new educational system for the colony. The astronomical work involved surveying 3,000 areas of the southern sky, counting nearly 69,000 single stars and 4,000 multiple systems. The data reduction and analysis occupied Herschel for nearly a decade, following his return to England in 1838. The extraordinary *Cape Observations*, published in 1847, represents the scientific labor of a single individual, labor of a magnitude unlikely ever to be repeated.

The last decades of Herschel's life were devoted to public service, general scientific writing, his translation of the *Iliad* and, as always, the star catalogues. His final work, a catalogue of 10,300 double stars, appeared shortly after his death in 1871.

The *Preliminary Discourse* of 1830 stems from one of the most fruitful and varied periods of Herschel's scientific career. It was a period of prestigious awards and of personal scientific achievement. It was also just at the beginning of his life with Margaret, a time when Herschel was filled with enthusiasm for science and passion for life. This spirit of enthusiasm enlivens the pages of the *Discourse,* blending easily with what we now identify as the progressive faith of the Victorian era. The *Discourse* opens with an engraving of Francis Bacon (1561–1626), who is eulogized in sections 96–106 as the theoretician of a new scientific age. The *Discourse* closes with a chapter devoted to explaining the grounds for what Herschel sees as the more rapid advance of science in his time than in any other. The reader will find citations there from the popular essays of a musician, the English composer William Jackson (1730–1803). Herschel's father was also a composer and a music teacher before he turned astronomer, and so perhaps his father was on his mind as Herschel framed his own statement of the progressive faith in that concluding chapter, "In whatever state of knowledge we may conceive man to be placed, his progress towards a yet higher state need never fear a check, but must continue till the last existence of society" (section 392).

The *Discourse,* however, is more than a sustained ode to the progress of science and humankind. It is also a careful analysis and explication of what Herschel regards as the principles and methods of scientific investigation, told from the perspective of someone himself engaged successfully and actively in scientific work, both at the theoretical level and at the level of experiment or observation. The framework of Herschel's story is Baconian; that is, he seeks to show how scientific knowledge is grounded in empirical investigation, and how it relates to the practical affairs of society. But unlike Bacon, Herschel's attitude toward the theoretical and hypothetical aspects of science is unambiguously affirmative. Indeed in the course of articulating the basis for his own model of an empirically rooted hierarchy of laws and high-level theories, Herschel explains the elaborate feedback loop linking inductive generalization with the deductive testing of consequences, a conception that is familiar to us all these days as "the hypothetical-deductive method."

The *Discourse* is divided into three parts. In Part I Herschel seeks to justify the study of "natural philosophy" (i.e., science) on its own grounds against the charge that, as a theoretical activity, it tends to undermine religion and is therefore morally dangerous to pursue. He also defends against the charge that, as a practical activity, science is merely an instrument contributing to our "pampered appetites" (section 7) for the material things of life. To the contrary, using (in part) a curious argument from atomism (section 28), Herschel contends that science necessarily supports religion and subverts materialism. The general picture of science that emerges from the discussion is of an activity whose pursuit actually enhances our moral stature, and whose practical benefits Herschel ties up not just with a better standard of living but also with a democratization of society at large, since the "fruits of science are in their nature diffusive, and cannot be enjoyed in any exclusive manner by a few" (section 62). The reader familiar with the classical arguments for free expression, in terms of a marketplace of ideas, for example, as later expressed in John Stuart Mill's *On Lib-*

erty (1859), will find similar arguments deployed by Herschel against scientific elites and in aid of the public dissemination and demystification of scientific expertise (section 63). Indeed one of the most striking features of Herschel's Baconian setting of the scientific enterprise in Part I is his emphasis on science as a community activity, both as generated *by* a community of coworkers and as generated *for* the community at large. Thus Herschel takes very seriously his own characterization of science as "the knowledge of many, orderly and methodically digested and arranged so as to become attainable by one" (section 13). For Herschel, science is foremost a human and social enterprise.

Part II of the *Discourse* moves from this general conception of science to its epistemology and methods. Herschel's epistemological orientation is clearly toward the empiricist tradition inherited from Bacon; that is, Herschel takes for granted that experience is the "only intimate source of our knowledge of nature and its laws" (section 67). Moreover, given Herschel's emphasis on the social character of science, for him "experience" refers to the collective experience of mankind. Nevertheless there is also a significant rationalist side to Herschel's thinking, and of the sort often associated with Bacon's younger continental contemporary and methodological rival, René Descartes (1596–1650). Descartes had argued that each of us has an innate and natural capacity for knowledge, the so-called "light of nature" that, according to Descartes, was placed in us by God. The problem of acquiring knowledge of nature, then, was just that of how to free our minds so as not to interfere with the operation of this innate capacity. Thus

Descartes proposed his famous criterion, that whatever we conceive of "clearly and distinctly" must be so. From this rationalist perspective one could expect a continual growth in knowledge as a natural outcome of proceeding methodically and carefully, and keeping our minds clear. In Herschel (as, indeed, in Bacon as well) we find a blend of empiricism with a tendency toward this sort of rationalism. He writes, "To experience we refer, as the only ground of all physical enquiry. But before experience itself can be used with advantage, there is one preliminary step to make, which depends wholly on ourselves: it is the absolute dismissal and clearing the mind of all prejudice" (section 68; see also section 108). An important theme that links empiricism with rationalism is the idea that knowledge has a foundation (or "ground" as Herschel often puts it). To be sure, the empiricist foundation is in sense experience, whereas the rationalist one involves innate capacities of the mind (or "first principles" of thought). But whether from the bottom up or from the top down, the idea of a foundation motivates the concept of progress through the application of the scientific method, and leads one to think that the scientific method itself constitutes a coherent subject, one worthy of separate investigation apart from a mere survey of the various methods of the different sciences. Herschel simply presupposes this foundationalist blend of empiricism and rationalism without examination or defense. It is the background against which his account proceeds.

According to Herschel science originates in observation and experimentation, these being distinguished as more passive (observation) versus more active (experimentation). Herschel favors the active, and holds that science began really to advance only when investigators carefully started to intervene in nature experimentally, and to judge the results. The results are how things appear to us, the *phenomena,* and they are in general complex, requiring analysis into their constituent parts. Herschel gives as an illustration one of his favorite topics, the phenomenon of sound (section 79). The analysis has to do with how sound is produced, communicated and, finally, experienced. That is, what Herschel illustrates as an analysis of a phenomenon into constituent parts amounts to an explanation of the phenomenon in terms of causes and natural laws.

Herschel regards a law of nature as "a statement . . . of what will happen in certain general contingencies" (sections 80 and 89) or, equivalently, as a statement of an invariable connection (sections 91 and 92), and he takes causes (or causal factors) to be what grounds the truth of such law-statements. According to Herschel, the search for causes and laws is one of the prime goals of science and, as we have seen, thinking in terms of causes and laws is threaded throughout the whole scientific enterprise, right from its ground in observation and experiment. This can lead to problems, however, in reporting observations or experimental results insofar as extraneous causal or theoretical language may infect the reports, making it difficult to get at the fact of the matter (section 125). Thus Herschel's account of observational methods is delicately balanced between, on the one side, the recognition of the necessity for a theoretical orientation and, on the other, the recognition of the dangers with regard to objectivity that can result from such an orientation. There is a similar tension that Herschel brings out in his discussion of classification (Chapter V, Part II), which is the next step up from observation. Herschel sees clearly that the grouping of natural phenomena into similarity classes must employ knowledge of constituent causes and lawlike connections. At the same time, he is well aware that such classification is also necessary in order to establish any causal or lawlike connections at the outset. Herschel proffers no magical solution for these dilemmas of practice, except for a general counsel of care, an emphasis on quantitative methods where feasible, and good sense. In noticing the issues, however, Herschel saw farther than many of *his* contemporaries, and many of ours.

In Chapter VI of Part II Herschel articulates a set of general rules for determining causes. In this he follows Bacon and David Hume (Part III, Book I of the *Treatise*, section 15). Herschel's rules are the material from which Mill developed his own well-known five "methods" (or "canons") of induction, in *A System of Logic* (1843). They, in turn, were criticized by Herschel's friend Whewell in *Of Induction, With Especial Reference to Mr. J. Stuart Mill's System of Logic* (1849). Ironically, Whewell's criticism brings the discussion round full circle. Whewell criticizes Mill's methods because they simply take for granted a preliminary analysis of the phenomena, one which would itself require the use of inductive methods; i.e., Whewell criticizes Mill for ignoring the very circularity in the conduct of causal analysis that Herschel had seen so clearly. According to Herschel's account, the rules for determining causes were equally rules for getting at the low level laws and generalizations codifying the causal connections.

Once the laws were articulated in this inductive manner, Herschel sought to use them to explain further phenomena; that is, to explore the consequences of the laws deductively. In this regard he emphasizes "that it is very important to observe that the successful process of scientific enquiry demands continually the alternate use of *inductive* and *deductive* methods" (section 184). This is the feedback loop referred to earlier, by means of which Herschel connects the discovery of laws with their testing and verification.

We can summarize Herschel's conception of scientific method as follows. We begin with observation and experiment, using analogies to help sort the results into causally relevant classes (which he calls "general facts" [section 94]), and then by comparing these classes we arrive at low-level laws. Proceeding from this stage via the same methods of causal analysis, we move to more general groupings and then to laws of wider scope. This process repeats itself through many stages until we arrive at very broad scientific theories. Running in tandem, all along this inductive line, are feedback mechanisms which check and correct the classifications and putative laws at each stage. They involve deductions of the consequences and their comparison with new observational data. As the inductive hierarchy builds up, Herschel notes that there is a natural tendency for the labor to divide between experimental work and work that is of a more theoretical kind (section 126). Thus Herschel's inductive scheme suggests an explanation for the division of labor that we see in science.

In his discussion of "residual phenomena" (sections 158–61), Herschel recognizes that the comparison of the law or theory with the data may be inconclusive insofar as one fails to account for all the observed phenomena. When this occurs, we may simply have to shelve what is unexplained, hoping to come back to it at a later stage of development. Similarly, Herschel recognizes that even when many consequences of a theory are well-confirmed, it may still be prudent to suspend judgment concerning the finer details of the theoretical story so that even confirmation may be inconclusive (section 216). However, Herschel does not treat the question of exactly how a "correction" is made when the deductive consequences of a law or theory do not fit the observations. Hence he misses one of the central problems of twentieth-century philosophy of science, the Poincaré/Duhem problem concerning falsification and the introduction of ad hoc hypotheses. Nevertheless, Herschel is quite sensitive to the inconclusive character of the self-correcting mechanisms that he does discuss. In fact, his neo-Baconian scheme for arriving at theories inductively is a rather looser and more open-ended scheme than talk of induction and "the scientific method" might suggest.

In Chapter VII of Part II the open-ended character of science emerges even more clearly. For there Herschel is concerned with hypotheses and theories that postulate "hidden processes" in order to explain and analyze phenomena, for example, theories about the nature of radiation (e.g., heat or light). In section 210 Herschel recognizes that although we may arrive at such theories by means of the articulated stages of induction, we may just as legitimately arrive there simply "by forming at once a bold hypothesis." All that matters is that we test the hypothesis by pursuing its empirical consequences to the finest details.

Thus the justification of a hypothesis need not always depend on whether its discovery has an inductive pedigree although, Herschel adds, "There is no doubt, however, that the safest course, when it can be followed, is to rise by induction[s] . . ." (section 217).

The reference to a "safest course" in this passage reminds us that the goal in a scientific investigation, as Herschel conceives it, is to discover the truth of the matter. Can we, then, take a well-confirmed and tested theory that invokes hidden processes as true? Do our methods of verification, which amount to checking the empirical consequences, justify us in believing that the postulated hidden structure is real? Herschel hedges his answer to these questions. On the one hand he says that, given the weight of evidence, we "cannot refuse to admit" such a theory. On the other hand, perhaps it would be prudent to admit it as only truth-*like*, which is to say that we do not affirm its literal truth but merely accept it as empirically adequate (section 220). Herschel's emphasis here is on the utility of such hypotheses and theories at the empirical level, where they may provide a reliable scaffolding for the organization of data and low-level laws. His worry is that we may "quite mistake the scaffold for the pile" (section 216). Thus with regard to the reality of hidden entities, Herschel's attitude is an instrumentalist one: If well confirmed, accept the theory that postulates them as a reliable guide to the phenomena, but do not necessarily believe it. This instrumentalism, however, does not sit easily with Herschel's faith in the continued progress of science toward its goal of finding the truth about nature. In the very last sentence of the *Discourse*, as though to address this unease, Herschel invokes simplicity as the criterion "on which the mind rests satisfied that it has attained the truth."

Part III of the *Discourse* turns from philosophy of science to history of science. In these chapters Herschel illustrates his methodological precepts by sketching the development of the physical sciences of his day. No doubt these chapters were also intended as appetizers for the more detailed lessons in science that were to follow in future volumes of the *Cabinet Cyclopedia.* It must be said that some of Herschel's historical forays in Part III are appalling. This is especially true of Chapter III, on astronomy. There Herschel begins with the familiar Baconian tirade against Aristotle, as "impeding progress," and then proceeds to tell the subsequent history as though it were a case of pure induction from the data, unfettered by hypotheses or theories, and as though the whole historical record led in a simple straight line directly to Newton's celestial mechanics. By contrast, Herschel's discussion of the wave versus the particle theory of light, in Chapter II, Part III is sensitive and subtle. Herschel himself had struggled through the competing experiments and theoretical refinements in this area, coming down, in the end, on the side of the wave theory. He retells this story brilliantly, and even follows his own instrumentalist precepts by recognizing that although the weight of evidence does seem to favor the wave theory we should nevertheless be wary of accepting that as the truth of the matter. In view of the revival of a particle theory in Einstein's light quantum hypothesis of 1905, and the subsequent history of the so-called wave/particle duality, Herschel's caution here seems well placed. The *Discourse* concludes with a chapter reflecting on the mechanisms that Herschel sees as aiding the rapid progress of science. In this discussion he returns to his emphasis on community aspects of science: the role of public education, periodicals and scientific societies, and the general interplay between science and society at large.

This brief introduction to Herschel's philosophy of science has emphasized some of his insightful contributions to the subject, as well as some of the uneasy ways in which Herschel's ideas straddle contrasting positions. For a scientist, insightful hedging may be the natural attitude toward science. This was certainly the case with Einstein, whose description of the scientist as philosopher might well have taken Herschel's *Discourse* as its model:

> (The scientist) must appear to the systematic epistemologist as a type of unscrupulous opportunist. He appears as *realist* insofar as he seeks to describe a world independent of the acts of perception; as *idealist* insofar as he looks upon the concepts and theories as the free inventions of the human spirit (not logically derivable from what is empirically given); as *positivist* insofar as he considers his concepts and theories justified *only* to the extent to which they furnish a logical representation of relations among sensory experiences. He may even appear as *Platonist* or *Pythagorean* insofar as he considers the viewpoint of logical simplicity as an indispensable and effective tool of his research. (A. Einstein, *Albert Einstein: Philosopher-Scientist,* ed. P.A. Shilpp [Open Court: LaSalle Illinois, 1949], p. 684.)

Arthur Fine

PRELIMINARY DISCOURSE on the Study of NATURAL PHILOSOPHY

JOHN FREDERICK WILLIAM HERSCHEL, ESQ. A.M.

LATE FELLOW OF ST JOHN'S COLLEGE, CAMBRIDGE. &c. &c. &c.

H. Corbould del. *E. Finden sculp.*

NATURÆ MINISTER ET INTERPRES

London:

PRINTED FOR LONGMAN, REES, ORME, BROWN, & GREEN: PATERNOSTER ROW

AND JOHN TAYLOR, UPPER GOWER STREET

1830.

"HOMO, NATURÆ MINISTER ET INTERPRES, TANTUM FACIT ET
"INTELLIGIT QUANTUM DE NATURÆ ORDINE RE VEL MENTE OB-
"SERVAVERIT: NEC AMPLIUS SCIT AUT POTEST."

BACON, NOVUM ORGANUM, APHOR. 1.

MAN, AS THE MINISTER AND INTERPRETER OF NATURE, IS LIMITED IN ACT AND UNDERSTANDING BY HIS OBSERVATION OF THE ORDER OF NATURE: NEITHER HIS KNOWLEDGE NOR HIS POWER EXTENDS FARTHER.

CONTENTS.

PART I.

OF THE GENERAL NATURE AND ADVANTAGES OF THE STUDY OF THE PHYSICAL SCIENCES.

CHAP. I.

CHAP. II.

CHAP. III.

PART II.

OF THE PRINCIPLES ON WHICH PHYSICAL SCIENCE RELIES FOR ITS SUCCESSFUL PROSECUTION, AND THE RULES BY WHICH A SYSTEMATIC EXAMINATION OF NATURE SHOULD BE CONDUCTED, WITH ILLUSTRATIONS OF THEIR INFLUENCE AS EXEMPLIFIED IN THE HISTORY OF ITS PROGRESS.

PART III.

OF THE SUBDIVISION OF PHYSICS INTO DISTINCT BRANCHES, AND THEIR MUTUAL RELATIONS.

" In primis, hominis est propria VERI inquisitio atque investigatio. Itaque
" cum sumus negotiis necessariis, curisque vacui, tum avemus aliquid
" videre, audire, ac dicere, cognitionemque rerum, aut occultarum aut ad-
" mirabilium, ad benè beatéque vivendum necessariam ducimus; — ex quo
" intelligitur, quod VERUM, simplex, sincerumque sit, id esse naturæ hominis
" aptissimum. Huic veri videndi cupiditati adjuncta est appetitio quæ-
" dam principatûs, ut nemini parere animus benè a naturâ informatus velit,
" nisi præcipienti, aut docenti, aut utilitatis causâ justè et legitimè impe-
" ranti: ex quo animi magnitudo existit, et humanarum rerum contemtio."

CICERO, DE OFFICIIS, Lib. I. § 13.

Before all other things, man is distinguished by his pursuit and investigation of TRUTH. And hence, when free from needful business and cares, we delight to see, to hear, and to communicate, and consider a knowledge of many admirable and abstruse things necessary to the good conduct and happiness of our lives: whence it is clear that whatsoever is TRUE, simple, and direct, the same is most congenial to our nature as men. Closely allied with this earnest longing to see and know the truth, is a kind of dignified and princely sentiment which forbids a mind, naturally well constituted, to submit its faculties to any but those who announce it in precept or in doctrine, or to yield obedience to any orders but such as are at once just, lawful, and founded on utility. From this source spring greatness of mind and contempt of worldly advantages and troubles.

PRELIMINARY DISCOURSE

ON

THE STUDY

OF

NATURAL PHILOSOPHY.

PART I.

OF THE GENERAL NATURE AND ADVANTAGES OF THE STUDY OF THE PHYSICAL SCIENCES.

CHAPTER I.

OF MAN REGARDED AS A CREATURE OF INSTINCT, OF REASON, AND SPECULATION.—GENERAL INFLUENCE OF SCIENTIFIC PURSUITS ON THE MIND.

(1.) The situation of man on the globe he inhabits, and over which he has obtained the control, is in many respects exceedingly remarkable. Compared with its other denizens, he seems, if we regard only his physical constitution, in almost every respect their inferior, and equally unprovided for the supply of his natural wants and his defence against the innumerable enemies which surround him. No other animal passes so large a portion of its existence in a state of absolute helplessness, or falls in old age into such protracted and lamentable imbecility. To no other warm-blooded animal has nature denied that indispensable covering without which the vicissitudes of a temperate and the rigours of a cold climate are equally insupportable; and to scarcely any has she been so sparing in external weapons, whether for attack or defence. Destitute alike of speed to avoid and of arms to repel the aggressions of his voracious foes; tenderly susceptible of atmospheric influences; and unfitted for the coarse aliments which the earth affords spontaneously during at least two thirds of the year, even in temperate climates,—man, if abandoned to mere instinct, would be of all creatures the most destitute and miserable. Distracted by terror and goaded by famine; driven to the most abject expedients for concealment from his enemies, and to the most cowardly devices for the seizure and destruction of his nobler prey, his existence would be one continued subterfuge or stratagem;—his dwelling would be in dens of the earth, in clefts of rocks, or in the hollows of trees; his food worms, and the lower reptiles, or such few and crude productions of the soil as his organs could be brought to assimilate, varied with occasional relics, mangled by more powerful beasts of prey, or contemned by their more pampered choice. Remarkable only for the absence of those powers and qualities which obtain for other animals a degree of security and respect, he would be disregarded by some, and hunted down by others, till after a few generations his species would become altogether extinct, or, at best, would be restricted to a few islands in tropical regions, where the warmth of the climate, the paucity of enemies, and the abundance of vegetable food, might permit it to linger.

(2.) Yet man is the undisputed lord of the creation. The strongest and fiercest of his fellow-creatures, — the whale, the elephant, the eagle, and the tiger,—are slaughtered by him to supply his most capricious wants, or tamed to do him service, or imprisoned to make him sport. The spoils of all nature are in daily requisition for his most common uses, yielded with more or less readiness, or wrested with reluctance, from the mine, the forest, the ocean, and the air. Such are the first fruits of reason. Were they the only or the principal ones, were the mere acquisition of power over the materials, and the less gifted animals which surround us, and the consequent increase of our external comforts, and our means of preservation and sensual enjoyment, the sum of the privileges which the possession of this faculty conferred, we should after all have little to plume ourselves upon. But this is so far from being the case, that every one who passes his life in tolerable ease and comfort, or rather whose whole time is not anxiously consumed in providing the absolute necessaries of existence, is conscious of wants and cravings in which the senses have no part, of a series of pains and pleasures totally distinct in kind from any which the infliction of bodily misery or the gratification of bodily appetites has ever afforded him; and if he has experienced these pleasures and these pains in any degree of intensity, he will readily admit them to hold a much higher rank, and to deserve much more attention, than the former class. Independent of the pleasures of fancy and imagination, and social converse, man is constituted a speculative being; he contemplates the world, and the objects around him, not with a passive, indifferent gaze, as a set of phenomena in which he has no further interest than as they affect his immediate situation, and can be rendered subservient to his comfort, but as a system disposed with order and design. He approves and feels the highest admiration for the harmony of its parts, the skill and efficiency of its contrivances. Some of these which he can best trace and understand he attempts to imitate, and finds that to a certain extent, though rudely and imperfectly, he can succeed, — in others, that although he can comprehend the nature of the contrivance, he is totally destitute of all means of imitation; — while in others, again, and those evidently the most important, though he sees the effect produced, yet the means by which it is done are alike beyond his knowledge and his control. Thus he is led to the conception of a Power and an Intelligence superior to his own, and adequate to the production and maintenance of all that he sees in nature, — a Power and Intelligence to which he may well apply the term infinite, since he not only sees no actual limit to the instances in which they are manifested, but finds, on the contrary, that the farther he enquires, and the wider his sphere of observation extends, they continually open upon him in increasing abundance; and that as the study of one prepares him to understand and appreciate another, refinement follows on refinement, wonder on wonder, till his faculties become bewildered in admiration, and his intellect falls back on itself in utter hopelessness of arriving at an end.

(3.) When from external objects he turns his view upon himself, on his own vital and intellectual faculties, he finds that he possesses a power of examining and analysing his own nature to a certain extent, but no farther. In his corporeal frame he is sensible of a power to communicate a certain moderate amount of motion to himself and other objects; that this power depends on his will, and that its exertion can be suspended or increased at pleasure within certain limits; but *how* his will acts on his limbs he has no consciousness: and whence he derives the power he thus exercises, there is nothing to assure him, however he may long to know. His senses, too, inform him of a multitude of particulars respecting the external world, and he perceives an apparatus by which impressions from without may be transmitted, as a sort of signals to the interior of his person, and ultimately to his brain, wherein he is obscurely sensible that the thinking, feeling, reasoning being he calls *himself*, more especially resides; but by what means he becomes conscious of these impressions, and what is the nature of the immediate communication between that inward sentient being, and that machinery, his outward man, he has not the slightest conception.

(4.) Again, when he contemplates still more attentively the thoughts, acts, and passions of this his sentient intelligent self, he finds, indeed, that he can remember, and by the aid of memory can compare and discriminate, can judge and resolve, and, above all, that he is irresistibly impelled, from the perception of any phenomenon without or within him, to infer the existence of something prior which stands to it in the relation of a *cause*, without which it would not be, and that this knowledge of causes and their consequences is what, in almost every instance, determines his choice and will, in cases where he is nevertheless conscious of perfect freedom to act or not to act. He finds, too, that it is in his power to acquire more or less knowledge of causes and effects according to the degree of attention he bestows upon them, which attention is again in great measure a voluntary act; and often when his choice has been decided on imperfect knowledge or insufficient attention, he finds reason to correct his judgment, though perhaps too late to influence his decision by after consideration. A world within him is thus opened to his intellectual view, abounding with phenomena and relations, and of the highest immediate interest. But while he cannot help perceiving that the insight he is enabled to obtain into this internal sphere of thought and feeling is in reality the source of all his power, the very fountain of his predominance over external nature, he yet feels himself capable of entering only very imperfectly into these recesses of his own bosom, and analysing the operations of his mind,—in this as in all other things, in short, "*a being darkly wise*;" seeing that all the longest life and most vigorous intellect can give him power to discover by his own research, or time to know by availing himself of that of others, serves only to place him on the very frontier of knowledge, and afford a distant glimpse of boundless realms beyond, where no human thought has penetrated, but which yet he is sure must be no less familiarly known to that Intelligence which he traces throughout creation than the most obvious truths which he himself daily applies to his most trifling purposes. Is it wonderful that a being so constituted should first encourage a hope, and by degrees acknowledge an assurance, that his intellectual existence will not terminate with the dissolution of his corporeal frame, but rather that in a future state of being, disencumbered of a thousand obstructions which his present situation throws in his way, endowed with acuter senses, and higher faculties, he shall drink deep at that fountain of beneficent wisdom for which the slight taste obtained on earth has given him so keen a relish?

(5.) Nothing, then, can be more unfounded than the objection which has been taken, *in limine*, by persons, well meaning perhaps, certainly narrow-minded, against the study of natural philosophy, and indeed against all science, — that it fosters in its cultivators an undue and overweening self-conceit, leads them to doubt the immortality of the soul, and to scoff at revealed religion. Its natural effect, we may confidently assert, on every well constituted mind is and must be the direct contrary. No doubt, the testimony of natural reason, on whatever exercised, must of necessity stop short of those truths which it is the object of revelation to make known; but, while it places the existence and principal attributes of a Deity on such grounds as to render doubt absurd and atheism ridiculous, it unquestionably opposes no natural or necessary obstacle to further progress: on the contrary, by cherishing as a vital principle an unbounded spirit of enquiry, and ardency of expectation, it unfetters the mind from prejudices of every kind, and leaves it open and free to every impression of a higher nature which it is susceptible of receiving, guarding only against enthusiasm and self-deception by a habit of strict investigation, but encouraging, rather than suppressing, every thing that can offer a prospect or a hope beyond the present obscure and unsatisfactory state. The character of the true philosopher is to hope all things not impossible, and to believe all things not unreasonable. He who has seen obscurities which appeared impenetrable in physical and mathematical science suddenly dispelled, and the most barren and unpromising fields of enquiry converted, as if by inspiration, into rich and inexhaustible springs of knowledge and power on a simple change of our point of view, or by merely bringing to bear on them some principle which it never occurred before to try, will surely be the very last to acquiesce in any dispiriting prospects of either the present or future destinies of mankind; while, on the other hand, the boundless views of intellectual and moral as well as material relations which open on him on all hands in the course of these pursuits, the knowledge of the trivial place he occupies in the scale of creation, and the sense continually pressed upon him of his own weakness and incapacity to suspend or modify the slightest movement of the vast machinery he sees in action around him, must effectually convince him that humility of pretension, no less than confidence of hope, is what best becomes his character.

(6.) But while we thus vindicate the study of natural philosophy from a charge at one time formidable from the pertinacity and acrimony with which it was urged, and still occasionally brought forward to the distress and disgust of every well constituted mind, we must take care that the testimony afforded by science to religion, be its extent or value what it may, shall be at least independent, unbiassed, and spontaneous. We do not here allude to such reasoners as would make all nature bend to their narrow interpretations of obscure and difficult passages in the sacred writings: such a course might well become the persecutors of Galileo and the other bigots of the fifteenth and sixteenth centuries, but can only be adopted by dreamers in the present age. But, without going these lengths, it is no uncommon thing to find persons, earnestly attached to science and anxious for its promotion, who yet manifest a morbid sensibility on points of this kind, — who exult and applaud when any fact starts up explanatory (as they suppose) of some scriptural allusion, and who feel pained and disappointed when the general course of discovery in any department of science runs wide of the notions with which particular passages in the Bible may have impressed themselves. To persons of such a frame of mind it ought to suffice to remark, on the one hand, that truth can never be opposed to truth, and, on the other, that error is only to be effectually confounded by searching deep and tracing it to its source. Nevertheless, it were much to be wished that such persons, estimable and excellent as they for the most part are, before they throw the weight of their applause or discredit into the scale of scientific opinion on such grounds, would reflect, first, that the credit and respectability of *any* evidence may be destroyed by tampering with its *honesty* ; and, secondly, that this very disposition of mind implies a lurking mistrust in its own principles, since the grand and indeed only character of truth is its capability of enduring the test of universal experience, and coming unchanged out of every possible form of *fair* discussion.

(7.) But if science may be vilified by representing it as opposed to religion, or trammelled by mistaken notions of the danger of free enquiry, there is yet another mode by which it may be degraded from its native dignity, and that is by placing it in the light of a mere appendage to and caterer for our pampered appetites. The question "*cui bono*" to what practical end and advantage do your researches tend? is one which the speculative philosopher who loves knowledge for its own sake, and enjoys, as a rational being should enjoy, the mere contemplation of harmonious and mutually dependent truths, can seldom hear without a sense of humiliation. He feels that there is a lofty and disinterested pleasure in his speculations which ought to exempt them from such questioning; communicating as they do to his own mind the purest happiness (after the exercise of the benevolent and moral feelings) of which human nature is susceptible, and tending to the injury of no one, he might surely allege *this* as a sufficient and direct reply to those who, having themselves little capacity, and less relish for intellectual pursuits, are constantly repeating upon him this enquiry. But if he can bring himself to descend from this high but fair ground, and justify himself, his pursuits, and his pleasures in the eyes of those around him, he has only to point to the history of all science, where speculations apparently the most unprofitable have almost invariably been those from which the greatest practical applications have emanated. What, for instance, could be apparently more unprofitable than the dry speculations of the ancient geometers on the properties of the conic sections, or than the dreams of Kepler (as they would naturally appear to his contemporaries) about the numerical harmonies of the universe? Yet these are the steps by which we have risen to a knowledge of the elliptic motions of the planets and the law of gravitation, with all its splendid theoretical consequences, and its inestimable practical results. The ridicule attached to "*Swing-swangs*" in Hooke's time * did not prevent him from reviving the proposal of the *pendulum* as a standard of measure, since so effectually wrought into practice by the genius and perseverance of captain Kater; — nor did that which Boyle encountered in his researches on the elasticity and pressure of the air act as any obstacle to the train of discovery which terminated in the steam-engine. The dreams of the alchemists led them on in the path of experiment, and drew attention to the wonders of chemistry, while they brought their advocates (it must be admitted) to merited contempt and ruin. But in this case it was moral dereliction which gave to ridicule a weight and power not necessarily or naturally belonging to it: but among the alchemists

* Hooke's Posthumous Works. Lond. 1705. — p. 472. and p. 458.

were men of superior minds, who reasoned while they worked, and who, not content to grope always in the dark, and blunder on their object, sought carefully in the observed nature of their agents for guides in their pursuits;—to these we owe the creation of experimental philosophy.

(8.) Not that it is meant, by any thing above said, to assert that there is no such thing as a great or a little in speculative philosophy, or to place the solution of an enigma on a level with the developement of a law of nature, still less to adopt the homely definition of Smith *, that a philosopher is a person whose trade it is to do nothing, and speculate on every thing. The speculations of the natural philosopher, however remote they may for a time lead him from beaten tracks and every-day uses, being grounded in the realities of nature, have all, of necessity, a practical application,—nay more, such applications form the very criterions of their truth, they afford the readiest and completest verifications of his theories; — verifications which he will no more neglect to test them by than an arithmetician would omit to *prove* his sums, or a cautious geometer to try his general theorems by particular cases. †

* Wealth of Nations, book i. chap. i. p. 15.

† On this subject, we cannot forbear citing a passage from one of the most profound but at the same time popular writers of our time, on a subject unconnected it is true with our own, but bearing strongly on the point before us. " But, if science be manifestly incomplete, and yet of the highest importance, it would surely be most unwise to restrain enquiry, conducted on just principles, even where the immediate practical utility of it was not visible. In mathematics, chemistry, and every branch of natural philosophy, how many are the enquiries

(9.) After all, however, it must be confessed, that to minds unacquainted with science, and unused to consider the mutual dependencies of its various branches, there is something neither unnatural nor altogether blamable in the ready occurrence of this question of direct advantage. It requires some habit of abstraction, some penetration of the mind with a tincture of scientific enquiry, some conviction of the value of those estimable and treasured principles which lie concealed in the most common and homely facts, — some experience, in fine, of success in developing and placing them in evidence, announcing them in precise terms, and applying them to the explanation of other facts of a less familiar character, or to the accomplishment of some obviously useful purpose: — to cure the mind of this tendency to rush at once upon its object, to undervalue the means in over-estimation of the end, and while gazing too intently at the goal which alone it has been accustomed to desire, to lose sight of the richness and variety of the prospects that offer themselves on either hand on the road.

(10.) We must never forget that it is principles, not phenomena, — laws, not insulated independent

---

necessary for their improvement and completion, which taken separately, do not appear to lead to any specifically advantageous purpose! how many useful inventions, and how much valuable and improving knowledge, would have been lost, if a rational curiosity, and a mere love of information, had not generally been allowed to be a sufficient motive for the search after truth!" - - Malthus's Principles of Political Economy, p. 16.

facts,—which are the objects of enquiry to the natural philosopher. As truth is single. and consistent with itself, a principle may be as completely and as plainly elucidated by the most familiar and simple fact, as by the most imposing and uncommon phenomenon. The colours which glitter on a soap-bubble are the immediate consequence of a principle the most important from the variety of phenomena it explains, and the most beautiful, from its simplicity and compendious neatness, in the whole science of optics. If the nature of periodical colours can be made intelligible by the contemplation of such a trivial object, from that moment it becomes a noble instrument in the eye of correct judgment; and to blow a large, regular, and durable soap-bubble may become the serious and praiseworthy endeavour of a sage, while children stand round and scoff, or children of a larger growth hold up their hands in astonishment at such waste of time and trouble. To the natural philosopher there is no natural object unimportant or trifling. From the least of nature's works he may learn the greatest lessons. The fall of an apple to the ground may raise his thoughts to the laws which govern the revolutions of the planets in their orbits; or the situation of a pebble may afford him evidence of the state of the globe he inhabits, myriads of ages ago, before his species became its denizens.

(11.) And this is, in fact, one of the great sources of delight which the study of natural science imparts to its votaries. A mind which has once imbibed a taste for scientific enquiry, and has learnt the habit of applying its principles readily to the cases which occur, has within itself an inexhaustible source of pure and exciting contemplations: — one would think that Shakspeare had such a mind in view when he describes a contemplative man as finding

> "Tongues in trees — books in the running brooks —
> Sermons in stones — and good in every thing."

Accustomed to trace the operation of general causes, and the exemplification of general laws, in circumstances where the uninformed and unenquiring eye perceives neither novelty nor beauty, he walks in the midst of wonders: every object which falls in his way elucidates some principle, affords some instruction, and impresses him with a sense of harmony and order. Nor is it a mere passive pleasure which is thus communicated. A thousand questions are continually arising in his mind, a thousand subjects of enquiry presenting themselves, which keep his faculties in constant exercise, and his thoughts perpetually on the wing, so that lassitude is excluded from his life, and that craving after artificial excitement and dissipation of mind, which leads so many into frivolous, unworthy, and destructive pursuits, is altogether eradicated from his bosom.

(12.) It is not one of the least advantages of these pursuits, which, however, they possess in common with every class of intellectual pleasures, that they are altogether independent of external circumstances, and are to be enjoyed in every situation in which a man can be placed in life. The highest degrees of worldly prosperity are so far from being incompatible with them, that they supply additional advantages for their pursuit, and that sort of fresh and renewed relish which arises partly from the sense of contrast, partly from experience of the peculiar pre-eminence they possess over the pleasures of sense in their capability of unlimited increase and continual repetition without satiety or distaste. They may be enjoyed, too, in the intervals of the most active business; and the calm and dispassionate interest with which they fill the mind renders them a most delightful retreat from the agitations and dissensions of the world, and from the conflict of passions, prejudices, and interests in which the man of business finds himself continually involved. There is something in the contemplation of general laws which powerfully persuades us to merge individual feeling, and to commit ourselves unreservedly to their disposal; while the observation of the calm, energetic regularity of nature, the immense scale of her operations, and the certainty with which her ends are attained, tends, irresistibly, to tranquillize and re-assure the mind, and render it less accessible to repining, selfish, and turbulent emotions. And this it does, not by debasing our nature into weak compliances and abject submission to circumstances, but by filling us, as from an inward spring, with a sense of nobleness and power which enables us to rise superior to them; by showing us our strength and innate dignity, and by calling upon us for the exercise of those powers and faculties by which we are susceptible of the comprehension of so much greatness, and which form, as it were, a link between ourselves and the best and noblest benefactors of our species, with whom we hold communion in thoughts and participate in discoveries which have raised them above their fellow mortals, and brought them nearer to their Creator.

CHAP. II.

OF ABSTRACT SCIENCE AS A PREPARATION FOR THE STUDY OF PHYSICS. — A PROFOUND ACQUAINTANCE WITH IT NOT INDISPENSABLE FOR A CLEAR UNDERSTANDING OF PHYSICAL LAWS. — HOW A CONVICTION OF THEIR TRUTH MAY BE OBTAINED WITHOUT IT. — INSTANCES. — FURTHER DIVISION OF THE SUBJECT.

(13.) SCIENCE is the knowledge of many, orderly and methodically digested and arranged, so as to become attainable by one. The knowledge of reasons and their conclusions constitutes *abstract*, that of causes and their effects, and of the laws of nature, *natural science.*

(14.) Abstract science is independent of a system of nature,— of a creation,— of every thing, in short, except memory, thought, and reason. Its objects are, first, those primary existences and relations which we cannot even conceive not to *be*, such as space, time, number, order, &c.; and, secondly, those artificial forms, or symbols, which thought has the power of creating for itself at pleasure, and substituting as representatives, by the aid of memory, for combinations of those primary objects and of its own conceptions,—either to facilitate the act of reasoning respecting them, or as convenient deposits of its own conclusions, or for their communication to others. Such are, first, *language*, oral or written; its conventional forms, which constitute grammar, and the rules for its use in argument, in which consists the logic of the schools; secondly, *notation*, which, applied to *number*, is *arithmetic*,— and, to the more general relations of abstract quantity or order, is *algebra;* and, thirdly, that higher kind of logic, which teaches us to use our reason in the most advantageous manner for the discovery of truth; which points out the criterions by which we may be sure we have attained it; and which, by detecting the sources of error, and exposing the haunts where fallacies are apt to lurk, at once warns us of their danger, and shows us how to avoid them. This greater logic may be termed *rational* *; while, to that inferior department which is conversant with words alone, the epithet *verbal* † may, for distinction, be applied.

(15.) A certain moderate degree of acquaintance with abstract science is highly desirable to every one who would make any considerable progress in physics. As the universe exists in time and place; and as motion, velocity, quantity, number, and order, are main elements of our knowledge of external things and their changes, an acquaintance with these, abstractedly considered, (that is to say, independent of any consideration of the particular things moved, measured, counted, or arranged,) must evidently be a useful preparation for the more complex study of nature. But there is yet another recommendation of such sciences as a preparation for the study of natural philosophy. Their objects are so definite, and our notions of them so distinct, that we can reason about them with an assurance,

* Λογος, *ratio*, reason. † Λογος, *verbum*, a word.

that the words and signs used in our reasonings are full and true representatives of the things signified; and, consequently, that when we use language or signs in argument, we neither, by their use, introduce extraneous notions, nor exclude any part of the case before us from consideration. For example: the words space, square, circle, a hundred, &c., convey to the mind notions so complete in themselves, and so distinct from every thing else, that we are sure when we use them we know and have in view the whole of our own meaning. It is widely different with words expressing natural objects and mixed relations. Take, for instance, *iron*. Different persons attach very different ideas to this word. One who has never heard of magnetism has a widely different notion of *iron* from one in the contrary predicament. The vulgar, who regard this metal as incombustible, and the chemist, who sees it burn with the utmost fury, and who has other reasons for regarding it as one of the most combustible bodies in nature; — the poet, who uses it as an emblem of rigidity; and the smith and engineer, in whose hands it is plastic, and moulded like wax into every form; — the jailer, who prizes it as an obstruction, and the electrician, who sees in it only a channel of open communication by which that most impassable of obstacles, the air, may be traversed by his imprisoned fluid, have all different, and all imperfect, notions of the same word. The meaning of such a term is like a rainbow — every body sees a different one, and all maintain it to be the same. So it is with nearly all our terms of sense. Some are indefinite, as hard or soft, light or heavy (terms which were at one time the sources of innumerable mistakes and controversies); some excessively complex, as man, life, instinct. But, what is worst of all, some, nay most, have two or three meanings; sufficiently distinct from each other to make a proposition true in one sense and false in another, or even false altogether; yet not distinct enough to keep us from confounding them in the process by which we arrived at it, or to enable us immediately to recognise the fallacy when led to it by a train of reasoning, each step of which we *think* we have examined and approved. Surely those who thus attach two senses to one word, or superadd a new meaning to an old one, act as absurdly as colonists who distribute themselves over the world, naming every place they come to by the names of those they have left, till all distinctions of geographical nomenclature are confounded, and till we are unable to decide whether an occurrence stated to have happened at Windsor took place in Europe, America, or Australia.*

(16.) It is, in fact, in this double or incomplete sense of words that we must look for the origin of a very large portion of the errors into which we fall. Now, the study of the abstract sciences, such as arithmetic, geometry, algebra, &c., while they afford scope for the exercise of reasoning about objects that are, or, at least, may be conceived to be, external to us;

* It were much to be wished that navigators would be more cautious in laying themselves open to a similar censure. On looking hastily over a map of the world we see three Melville Islands, two King George's Sounds, and Cape Blancos innumerable.

yet, being free from these sources of error and mistake, accustom us to the strict use of language as an instrument of reason, and by familiarising us in our progress towards truth to walk uprightly and straightforward on firm ground, give us that proper and dignified carriage of mind which could never be acquired by having always to pick our steps among obstructions and loose fragments, or to steady them in the reeling tempest of conflicting meanings.

(17.) But there is yet another point of view under which some acquaintance with abstract science may be regarded as highly desirable in general education, if not indispensably necessary, to impress on us the distinction between strict and vague reasoning, to show us what demonstration really *is*, and to give us thereby a full and intimate sense of the nature and strength of the evidence on which our knowledge of the actual system of nature, and the laws of natural phenomena, rests. For this purpose, however, a very moderate acquaintance with the more elementary branches of mathematics may suffice. The chain is laid before us, and every link is submitted to our unreserved examination, if we have patience and inclination to enter on such detail. Hundreds have gone through it, and will continue to do so; but, for the generality of mankind, it is enough to satisfy themselves of the solidity and adamantine texture of its materials, and the unreserved exposure of its weakest, as well as its strongest, parts. If, however, we content ourselves with this general view of the matter, we must be content also to take on trust, that is, on the authority of those who have examined deeper, every conclusion which cannot be made apparent to our senses. Now, among these there are many so very surprising, indeed apparently so extravagant, that it is quite impossible for any enquiring mind to rest contented with a mere hearsay statement of them, — we feel irresistibly impelled to enquire further into their truth. What mere assertion will make any man believe, that in one second of time, in one beat of the pendulum of a clock, a ray of light travels over 192,000 miles, and would therefore perform the tour of the world in about the same time that it requires to wink with our eyelids, and in much less than a swift runner occupies in taking a single stride? What mortal can be made to believe, without demonstration, that the sun is almost a million times larger than the earth? and that, although so remote from us, that a cannon ball shot directly towards it, and maintaining its full speed, would be twenty years in reaching it, it yet affects the earth by its attraction in an inappreciable instant of time? — a closeness of union of which we can form but a feeble, and totally inadequate, idea, by comparing it to any material connection; since the communication of an impulse to such a distance, by any solid intermedium we are acquainted with, would require, not moments, but whole years. And when, with pain and difficulty we have strained our imagination to conceive a distance so vast, a force so intense and penetrating, if we are told that the one dwindles to an insensible point, and the other is unfelt at the nearest of the fixed stars, from the mere effect of their remoteness, while among those very stars are some whose actual splendour exceeds by many hundred times that of the sun itself, although we may not deny the truth of the assertion, we cannot but feel the keenest curiosity to know *how* such things were ever made out.

(18.) The foregoing are among those results of scientific research which, by their magnitude, seem to transcend our powers of conception. There are others, again, which, from their minuteness, would appear to elude the grasp of thought, much more of distinct and accurate measurement. Who would not ask for demonstration, when told that a gnat's wing, in its ordinary flight, beats many hundred times in a second? or that there exist animated and regularly organised beings, many thousands of whose bodies laid close together would not extend an inch? But what are these to the astonishing truths which modern optical enquiries have disclosed, which teach us that every point of a medium through which a ray of light passes is affected with a succession of periodical movements, regularly recurring at equal intervals, no less than five hundred millions of millions of times in a single second! that it is by such movements, communicated to the nerves of our eyes, that we see — nay more, that it is the *difference* in the frequency of their recurrence which affects us with the sense of the diversity of colour; that, for instance, in acquiring the sensation of redness our eyes are affected four hundred and eighty-two millions of millions of times; of yellowness, five hundred and forty-two millions of millions of times; and of violet, seven hundred and seven millions of millions of times per second.* Do not such things sound more like

* Young. Lectures on Nat. Phil. ii. 627. See also Phil. Trans. 1801–2.

the ravings of madmen, than the sober conclusions of people in their waking senses?

(19.) They are, nevertheless, conclusions to which any one may most certainly arrive, who will only be at the trouble of examining the chain of reasoning by which they have been deduced; but, in order to do this, something beyond the mere elements of abstract science is required. Waving, however, such instances as these, which, after all, are rather calculated to surprise and astound than for any other purpose, it must be observed that it is not possible to satisfy ourselves completely that we *have* arrived at a true statement of any law of nature, until, setting out from such statement, and making it a foundation of reasoning, we can show, by strict argument, that the facts observed must follow from it as necessary logical consequences, and *this*, not vaguely and generally, but with all possible precision in time, place, weight, and measure.

(20.) To do this, however, as we shall presently see, requires in many cases a degree of knowledge of mathematics and geometry altogether unattainable by the generality of mankind, who have not the leisure, even if they all had the capacity, to enter into such enquiries, some of which are indeed of that degree of difficulty that they can be only successfully prosecuted by persons who devote to them their whole attention, and make them the serious business of their lives. But there is scarcely any person of good ordinary understanding, however little exercised in abstract enquiries, who may not be readily made to comprehend at least the general train of reasoning by which any of the great truths of physics are deduced, and the essential bearings and connections of the several parts of natural philosophy. There are whole branches too and very extensive and important ones, to which mathematical reasoning has never been at all applied; such as chemistry, geology, and natural history in general, and many others, in which it plays a very subordinate part, and of which the essential principles, and the grounds of application to useful purposes, may be perfectly well understood by a student who possesses no more mathematical knowledge than the rules of arithmetic; so that no one need be deterred from the acquisition of knowledge, or even from active original research in such subjects, by a want of mathematical information. Even in those branches which, like astronomy, optics, and dynamics, are almost exclusively under the dominion of mathematics, and in which no effectual progress can be made without *some* acquaintance with geometry, the principal *results* may be perfectly understood without it. To one incapable of following out the intricacies of mathematical demonstration, the conviction afforded by verified predictions must stand in the place of that purer and more satisfactory reliance which a verification of every step in the process of reasoning can alone afford, since every one will acknowledge the validity of pretensions which he is in the daily habit of seeing brought to the test of practice.

(21.) Among the verifications of this practical kind which abound in every department of physics, there are none more imposing than the precise prediction of the greater phenomena of astronomy; none, certainly, which carry a broader conviction home to every mind from their notoriety and unequivocal character. The prediction of eclipses has accordingly from the earliest ages excited the admiration of mankind, and been one grand instrument by which their allegiance (so to speak) to natural science, and their respect for its professors, has been maintained; and though strangely abused in unenlightened ages by the supernatural pretensions of astrologers, the credence given even to their absurdities shows the force of this kind of evidence on men's minds. The predictions of astronomers are, however, now far too familiar to endanger the just equipoise of our judgment, since even the return of comets, true to their paths and exact to the hour of their appointment, has ceased to amaze, though it must ever delight all who have souls capable of being penetrated by such beautiful instances of accordance between theory and facts. But the age of mere wonder in such things is past, and men prefer being guided and enlightened, to being astonished and dazzled. Eclipses, comets, and the like, afford but rare and transient displays of the powers of calculation, and of the certainty of the principles on which it is grounded. A page of "lunar distances" from the Nautical Almanack is worth all the eclipses that have ever happened for inspiring this necessary confidence in the conclusions of science. That a man, by merely measuring the moon's apparent distance from a star with a little portable instrument held in his hand, and applied to his eye, even with so unstable a footing as the deck of a ship, shall say positively, within five miles, where he is, on a boundless ocean, cannot but appear to persons ignorant of physical astronomy an approach to the miraculous. Yet, the alternatives of life and death, wealth and ruin, are daily and hourly staked with perfect confidence on these marvellous computations, which might almost seem to have been devised on purpose to show how closely the extremes of speculative refinement and practical utility can be brought to approximate. We have before us an anecdote communicated to us by a naval officer *, distinguished for the extent and variety of his attainments, which shows how impressive such results may become in practice. He sailed from San Blas on the west coast of Mexico, and after a voyage of 8000 miles, occupying 89 days, arrived off Rio de Janeiro, having, in this interval, passed through the Pacific Ocean, rounded Cape Horn, and crossed the South Atlantic, without making any land, or even seeing a single sail, with the exception of an American whaler off Cape Horn. Arrived within a week's sail of Rio, he set seriously about determining, by lunar observations, the precise line of the ship's course and its situation in it at a determinate moment, and having ascertained this within from five to ten miles, ran the rest of the way by those more ready and compendious methods, known to navigators, which can be safely employed for short trips between one known point and another, but which cannot be trusted in long voyages, where the moon is the only sure guide. The rest of the tale we are enabled by his kindness to state in his own words: —" We steered towards Rio de Janeiro for some days after taking the lunars

* Captain Basil Hall, R. N.

above described, and having arrived within fifteen or twenty miles of the coast, I hove to at four in the morning till the day should break, and then bore up; for although it was very hazy, we could see before us a couple of miles or so. About eight o'clock it became so foggy that I did not like to stand in farther, and was just bringing the ship to the wind again before sending the people to breakfast, when it suddenly cleared off, and I had the satisfaction of seeing the great Sugar Loaf Rock, which stands on one side of the harbour's mouth, so nearly right ahead that we had not to alter our course above a point in order to hit the entrance of Rio. This was the first land we had seen for three months, after crossing so many seas and being set backwards and forwards by innumerable currents and foul winds." The effect on all on board might well be conceived to have been electric; and it is needless to remark how essentially the authority of a commanding officer over his crew may be strengthened by the occurrence of such incidents, indicative of a degree of knowledge and consequent power beyond their reach.

(22.) But even such results as these, striking as they are, yet fall short of the force with which conviction is urged upon us when, through the medium of reasoning too abstract for common apprehension, we arrive at conclusions which outrun experience, and describe beforehand what will happen under new combinations, or even correct imperfect experiments, and lead us to a knowledge of facts contrary to received analogies drawn from an experience wrongly interpreted or overhastily generalised. To give an example: — every body knows that objects viewed through a transparent medium, such as water or glass, appear distorted or displaced. Thus, a stick in water appears bent, and an object seen through a prism or wedge of glass seems to be thrown aside from its true place. This effect is owing to what is called the *refraction* of light; and a simple rule discovered by Willebrod Snell enables any one to say exactly *how much* the stick will be bent, and *how far*, and in what *direction*, the apparent situation of an object seen through the glass will deviate from the real one. If a shilling be laid at the bottom of a basin of water and viewed obliquely, it will appear to be raised by the water; if instead of water spirits of wine be used it will appear more raised; if oil, still more: — but in none of these cases will it appear to be thrown *aside* to the *right* or *left* of its true place, however the eye be situated. The *plane*, in which are contained the eye, the object, and the point in the surface of the liquid at which the object is seen, is an upright or *vertical* plane; and this is one of the principal characters in the *ordinary refraction* of light, viz. that the ray by which we see an object through a refracting surface, although it undergoes a bending, and is, as it were, broken at the surface, yet, in pursuing its course to the eye, does not *quit a plane perpendicular to the refracting surface.* But there are again other substances, such as rock-crystal, and especially Iceland spar, which possess the singular property of *doubling* the image or appearance of an object seen through them in certain directions; so that instead of seeing one object we see two, side by side, when such a crystal or spar is interposed between the object and the eye; and if a ray or small sunbeam be thrown upon a surface of either of these substances, it will be split into two, making an angle with each other, and each pursuing its own separate course, — this is called *double refraction.* Now, of these images or doubly refracted rays, *one* always follows the same rule as if the substance were glass or water: its deviation can be correctly calculated by Snell's law above mentioned, and it does not quit the plane perpendicular to the refracting surface. The other ray, on the contrary, (which is therefore said to have undergone *extraordinary refraction*) *does* quit that plane, and the amount of its deviation from its former course requires for its determination a much more complicated rule, which cannot be understood or even stated without a pretty intimate knowledge of geometry. Now, rock-crystal and Iceland spar differ from glass in a very remarkable circumstance. They affect naturally certain regular figures, not being found in shapeless lumps, but in determinate geometrical forms; and they are susceptible of being cleft or split much easier in certain directions than in others — they have a *grain* which glass has not. When other substances having this peculiarity (and which are called *crystallized* substances) were examined, they were all, or by far the greater part, found to possess this singular property of *double refraction;* and it was very natural to conclude, therefore, that the same thing took place in all of them, viz. that of the two rays, into which any beam of light falling on the surface of such a substance was split, or of the two images of an object seen through it, *one* only was turned aside out of its *plane* and *extraordinarily* refracted, while the other followed the *ordinary* rule. Accordingly this was supposed to be the case; and not only so, but from some trials and measurements purposely made by a philosopher of great eminence, it was considered to be a fact sufficiently established by experiment.

(23.) Perhaps we might have remained long under this impression, for the measurements are delicate, and the subject very difficult. But it has lately been demonstrated by an eminent French philosopher and mathematician, M. Fresnel, that, granting certain *principles* or postulates, all the phenomena of double refraction, including perhaps the greatest variety of facts that have ever yet been arranged under one general head, may be satisfactorily explained and deduced from them by strict mathematical calculation; and *that*, when applied to the cases first mentioned, these principles give a satisfactory account of the *want* of the extraordinary image; *that* when applied to such cases as those of rock-crystal or Iceland spar, they also give a correct account of both the images, and agree in their conclusions with the rules before ascertained for them: but so far from coinciding with that part of the previous statement, which would make these conclusions extend to all crystallised substances, M. Fresnel's principles lead to a conclusion quite opposite, and point to a *fact* which had never been observed, viz. that in by far the greater number of crystallized substances which possess the property of double refraction, *neither* of the images follows the ordinary law, but both undergo a deviation from their original plane. Now this had never been observed to be the case in any previous trial, and all opinion was against it. But when put to the test of experiment in a great variety of new and ingenious methods, it was found to be fully verified; and to complete the evidence, the substances on whose imperfect examination the first erroneous conclusion was founded, having been lately subjected to a fresh and more scrupulous examination, the result has shown the insufficiency of the former measurements, and proved in perfect accordance with the newly discovered laws. Now it will be observed in this case, first, that, so far from the principles assumed by M. Fresnel being at all obvious, they are extremely remote from ordinary observation; and, secondly, that the chain of reasoning by which they are brought to the test is one of such length and complexity, and the purely mathematical difficulty of their application so great, that no *mere* good common sense, no general tact or ordinary practical reasoning, would afford the slightest chance of threading their mazes. Cases like this are the triumph of theories. They show at once how large a part pure reason has to perform in our examination of nature, and how implicit our reliance ought to be on that powerful and methodical system of rules and processes which constitute the modern mathematical analysis, in all the more difficult applications of exact calculation to her phenomena.

(24.) To take an instance more within ordinary apprehension. An eminent living geometer had proved by calculations, founded on strict optical principles, that in the *centre of the shadow* of a small circular plate of metal, exposed in a dark room to a beam of light emanating from a *very small brilliant point*, there ought to be no darkness, — in fact, *no shadow* at that place; but, on the contrary, a degree of illumination precisely as bright as if the metal plate were away. Strange and even impossible as this conclusion may seem, it has been put to the trial, and found perfectly correct. *

(25.) We shall now proceed to consider more particularly, and in detail, —

I. The nature and objects immediate and collateral of physical science, as regarded in itself, and in its application to the practical purposes of life, and its influence on the well-being and progress of society.

II. The principles on which it relies for its successful prosecution, and the rules by which a systematic examination of nature should be conducted, with examples illustrative of their influence.

III. The subdivision of physical science into distinct branches, and their mutual relations.

* We must caution our readers who would assure themselves of it by trial, that it is an experiment of some delicacy, and not to be made without several precautions to ensure success. For these we must refer to our original authority (Fresnel. Mémoire sur la Diffraction de la Lumiere, p. 124.); and the principles on which they depend will of course be detailed in that volume of the Cabinet Cyclopædia which is devoted to the subject of LIGHT.

CHAP. III.

OF THE NATURE AND OBJECTS, IMMEDIATE AND COLLATERAL, OF PHYSICAL SCIENCE, AS REGARDED IN ITSELF, AND IN ITS APPLICATION TO THE PRACTICAL PURPOSES OF LIFE, AND ITS INFLUENCE ON THE WELL-BEING AND PROGRESS OF SOCIETY.

(26.) THE first thing impressed on us from our earliest infancy is, that events do not succeed one another at random, but with a certain degree of order, regularity, and connection; — some constantly, and, as we are apt to think, immutably, — as the alternation of day and night, summer and winter,— others contingently, as the motion of a body from its place, if pushed, or the burning of a stick if thrust into the fire. The knowledge that the former class of events *has* gone on, uninterruptedly, for ages beyond all memory, impresses us with a strong expectation that it will continue to do so in the same manner; and thus our notion of an *order of nature* is originated and confirmed. If every thing were equally regular and periodical, and the succession of events liable to no change depending on our own will, it may be doubted whether we should ever think of looking for causes. No one regards the night as the cause of the day, or the day of night. They are alternate effects of a common cause, which their regular succession alone gives us no sufficient clue for determining. It is chiefly, perhaps entirely, from the other or contingent class of events that we gain our notions of cause and effect. From them alone we gather that there are such things as laws of nature. The very idea of a law includes that of contingency. "*Si quis mala carmina condidisset, fuste ferito;*" if such a case arise, such a course shall be followed, — if the match be applied to the gunpowder, it will explode. Every law is a provision for cases which *may* occur, and has relation to an infinite number of cases that never have occurred, and never will. Now, it is this provision, *à priori,* for contingencies, this contemplation of possible occurrences, and predisposal of what shall happen, that impresses us with the notion of a *law* and a *cause.* Among all the possible combinations of the fifty or sixty elements which chemistry shows to exist on the earth, it is likely, nay almost certain, that *some* have never been formed; that some elements, in some proportions, and under some circumstances, have never yet been placed in relation with one another. Yet no chemist can doubt that it is *already fixed* what they will do when the case does occur. They will obey certain laws, of which we know nothing at present, but which must *be* already fixed, or they could not be laws. It is not by habit, or by trial and failure, that they will learn what to do. When the contingency occurs, there will be no hesitation, no consultation; — their course will at once be decided, and will always be the same if it occur ever so often in succession, or in ever so many places at one and the same instant. This is the perfection of a law, that it includes all possible contingencies, and ensures implicit obedience, —and of this kind are the laws of nature.

(27.) This use of the word *law*, however, our readers will of course perceive has relation to us as understanding, rather than to the materials of which the universe consists as obeying, certain rules. To *obey* a law, to act in *compliance* with a rule, supposes an understanding and a will, a power of complying or not, in the being who obeys and complies, which we do not admit as belonging to mere matter. The Divine Author of the universe cannot be supposed to have laid down particular laws, enumerating all individual contingencies, which his materials have understood and obey,—this would be to attribute to him the imperfections of human legislation;—but rather, by creating them, endued with certain fixed qualities and powers, he has impressed them in their origin with the *spirit*, not the *letter*, of his law, and made all their subsequent combinations and relations inevitable consequences of this first impression, by which, however, we would no way be understood to deny the constant exercise of his direct power in maintaining the system of nature, or the ultimate emanation of every energy which material agents exert from his immediate will, acting in conformity with his own laws.

(28.) The discoveries of modern chemistry have gone far to establish the truth of an opinion entertained by some of the ancients, that the universe consists of distinct, separate, indivisible *atoms*, or individual beings so minute as to escape our senses, except when united by millions, and by this aggregation making up bodies of even the smallest visible bulk; and we have the strongest evidence that, although there exist great and essential differences in individuals among these atoms, they may yet all be arranged in a very limited number of groups or classes, all the individuals of each of which are, to all intents and purposes, *exactly alike* in all their properties. Now, when we see a great number of things precisely alike, we do not believe this similarity to have originated except from a common principle independent of them; and that we recognise this likeness, chiefly by the identity of their deportment under similar circumstances, strengthens rather than weakens the conclusion. A line of spinning-jennies *, or a regiment of soldiers dressed exactly alike, and going through precisely the same evolutions, gives us no idea of independent existence: we must see them act out of concert before we can believe them to have independent wills and properties, not impressed on them from without. And this conclusion, which would be strong even were there only two individuals precisely alike in *all* respects and *for ever*, acquires irresistible force when their number is multiplied beyond the power of imagination to conceive. If we mistake not, then, the discoveries alluded to effectually destroy the idea of an *eternal self-existent matter*, by giving to each of its atoms the essential characters, at once, of a *manufactured article*, and a *subordinate agent*.

(29.) But to ascend to the origin of things, and speculate on the creation, is not the business of the natural philosopher. An humbler field is sufficient

* Little reels used in cotton mills to twist the thread.

for him in the endeavour to discover, as far as our faculties will permit, what *are* these primary qualities originally and unalterably impressed on matter, and to discover the *spirit* of the laws of nature, which includes groups and classes of relations and facts from the *letter* which, as before observed, is presented to us by single phenomena: or if, after all, this should prove impossible; if such a step be beyond our faculties; and the essential qualities of material agents be really *occult*, or incapable of being expressed in any form intelligible to our understandings, at least to approach as near to their comprehension as the nature of the case will allow; and devise such forms of words as shall include and *represent* the greatest possible multitude and variety of phenomena.

(30.) Now, in this research there would seem one great question to be disposed of before our enquiries can even be commenced with any thing like a prospect of success, which is, whether the laws of nature themselves *have* that degree of permanence and fixity which can render them subjects of systematic discussion; or whether, on the other hand, the qualities of natural agents are subject to mutation from the lapse of time. To the ancients, who lived in the infancy of the world, or rather, in the infancy of man's experience, this was a very rational subject of question, and hence their distinctions between corruptible and incorruptible matter. Thus, according to some among them, the matter only of the celestial spaces is pure, immutable, and incorruptible, while all sublunary things are in a constant state of lapse and change; the world becoming paralysed and effete with age, and man himself deteriorating in character, and diminishing at once in intellectual and bodily stature. But to us, who have the experience of some additional thousands of years, the question of permanence is already, in a great measure, decided in the affirmative. The refined speculations of modern astronomy, grounding their conclusions on observations made at very remote periods, have proved to demonstration, that one at least of the great powers of nature, the force of gravitation, the main bond and support of the material universe, has undergone no change in intensity from a high antiquity. The stature of mankind is just what it was three thousand years ago, as the specimens of mummies which have been examined at various times sufficiently show. The intellect of Newton, Laplace, or La Grange, may stand in fair competition with that of Archimedes, Aristotle, or Plato; and the virtues and patriotism of Washington with the brightest examples of ancient history.

(31.) Again, the researches of chemists have shown that what the vulgar call corruption, destruction, &c., is nothing but a change of arrangement of the same ingredient elements, the disposition of the same materials into other forms, without the loss or actual destruction of a single atom; and thus any doubts of the permanence of natural laws are discountenanced, and the whole weight of *appearances* thrown into the opposite scale. One of the most obvious cases of apparent destruction is, when any thing is ground to dust and scattered to the winds. But it is one thing to grind a fabric to powder, and another to annihilate its materials: scattered as they may be, they must fall somewhere, and continue, if only as ingredients of the soil, to perform their humble but useful part in the economy of nature. The destruction produced by fire is more striking: in many cases, as in the burning of a piece of charcoal or a taper, there is no smoke, nothing visibly dissipated and carried away; the burning body wastes and disappears, while nothing *seems* to be produced but warmth and light, which we are not in the habit of considering as substances; and when all has disappeared, except perhaps some trifling ashes, we naturally enough suppose it is gone, lost, destroyed. But when the question is examined more exactly, we detect, in the invisible stream of heated air which ascends from the glowing coal or flaming wax, the *whole* ponderable matter, only united in a new combination with the air, and dissolved in it. Yet, so far from being thereby destroyed, it is only become again what it was before it existed in the form of charcoal or wax, an active agent in the business of the world, and a main support of vegetable and animal life, and is still susceptible of running again and again the same round, as circumstances may determine; so that, for aught we can see to the contrary, the same identical atom may lie concealed for thousands of centuries in a limestone rock; may at length be quarried, set free in the limekiln, mix with the air, be absorbed from it by plants, and, in succession, become a part of the frames of myriads of living beings, till some concurrence of events consigns it once more to a long repose, which, however, no way unfits it from again resuming its former activity.

(32.) Now, this absolute indestructibility of the ultimate materials of the world, in periods commensurate to our experience, and their obstinate retention of the same properties, under whatever variety of circumstances we choose to place them, however violent and seemingly contradictory to their natures, is, of itself, enough to render it highly improbable that time alone should have any influence over them. All that age or decay can do seems to be included in a wasting of parts which are only dissipated, not destroyed, or in a change of sensible properties, which chemistry demonstrates to arise only from new combinations of the same ingredients. But, after all, the question is one entirely of experience: we cannot be sure, *à priori*, that the laws of nature are *immutable;* but we can ascertain, by enquiry, *whether they change or not;* and to this enquiry all experience answers in the negative. It is not, of course, intended here to deny that great operations, productive of extensive changes in the visible state of nature, — such as, for instance, those contemplated by the geologists, and embracing for their completion vast periods of time, — are constantly going on; but these are consequences and fulfilments of the laws of nature, not contradictions or exceptions to them. No theorist regards such changes as alterations in the fundamental principles of nature; he only endeavours to reconcile them, and show how they result from laws already known, and judges of the correctness of his theory by their ultimate agreement.

(33.) But the laws of nature are not only permanent, but consistent, intelligible, and discoverable with such a moderate degree of research, as is calculated rather to stimulate than to weary curiosity. If we were set down, as creatures of another world, in any existing society of mankind, and began to speculate on their actions, we should find it difficult at first to ascertain whether they were subject to any laws at all: but when, by degrees, we had found out that they did consider themselves to be so; and would then proceed to ascertain, from their conduct and its consequences, what these laws were, and in what spirit conceived; though we might not perhaps have much difficulty in discovering single rules applicable to particular cases, yet, the moment we came to generalize, and endeavour from these to ascend, step by step, and discover any steady pervading principle, the mass of incongruities, absurdities, and contradictions, we should encounter, would either dishearten us from further enquiry, or satisfy us that what we were in search of did not exist. It is quite the contrary in nature; there we find no contradictions, no incongruities, but all is harmony. What once is learnt we never have to unlearn. As rules advance in generality, apparent exceptions become regular; and equivoque, in her sublime legislation, is as unheard of as maladministration.

(34.) Living, then, in a world where such laws obtain, and under their immediate dominion, it is manifestly of the utmost importance to know them, were it for no other reason than to be sure, in all we undertake, to have, at least, the law on our side, so as not to struggle in vain against some insuperable difficulty opposed to us by natural causes.

What pains and expense would not the alchemists, for instance, have been spared by a knowledge of those simple laws of composition and decomposition, which now preclude all idea of the attainment of their declared object! what an amount of ingenuity, thrown away on the pursuit of the perpetual motion, might have been turned to better use, if the simplest laws of mechanics had been known and attended to by the inventors of innumerable contrivances destined to that end! What tortures, inflicted on patients by imaginary cures of incurable diseases, might have been dispensed with, had a few simple principles of physiology been earlier recognised!

(35.) But if the laws of nature, on the one hand, are invincible opponents, on the other, they are irresistible auxiliaries; and it will not be amiss if we regard them in each of those characters, and consider the great importance of a knowledge of them to mankind,—

I. *In showing us how to avoid attempting impossibilities.*

II. *In securing us from important mistakes in attempting what is, in itself, possible, by means either inadequate, or actually opposed, to the end in view.*

III. *In enabling us to accomplish our ends in the easiest, shortest, most economical, and most effectual manner.*

IV. *In inducing us to attempt, and enabling us to accomplish, objects which, but for such knowledge, we should never have thought of undertaking.*

We shall therefore proceed to illustrate by examples the effect of physical knowledge under each of these heads: —

(36.) Ex. 1. (35.) I. It is not many years since an attempt was made to establish a colliery at Bexhill, in Sussex. The appearance of thin seams and sheets of fossil-wood and wood-coal, with some other indications similar to what occur in the neighbourhood of the great coal-beds in the north of England, having led to the sinking of a shaft, and the erection of machinery on a scale of vast expense, not less than eighty thousand pounds are said to have been laid out on this project, which, it is almost needless to add, proved completely abortive, as every geologist would have at once declared it must, the whole assemblage of geological facts being adverse to the existence of a regular coal-bed *in* the Hastings' *sand;* while this, on which Bexhill is situated, is separated from the *coal-strata* by a series of interposed beds of such enormous thickness as to render all idea of penetrating *through* them absurd. The history of mining operations is full of similar cases, where a very moderate acquaintance with the *usual order of nature,* to say nothing of theoretical views, would have saved many a sanguine adventurer from utter ruin.

(37.) Ex. 2. (35.) II. The smelting of iron requires the application of the most violent heat that can be raised, and is commonly performed in tall furnaces, urged by great iron bellows driven by steam-engines. Instead of employing this power to force *air* into the furnace through the intervention of bellows, it was, on one occasion, attempted to employ the steam itself in, apparently, a much less circuitous manner; viz. by directing the current of steam in a violent blast, from the boiler at once into the fire. From one of the known ingredients of steam being a highly inflammable body, and the other that essential part of the air which supports combustion, it was imagined that this would have the effect of increasing the fire to tenfold fury, whereas it simply *blew it out;* a result which a slight consideration of the laws of chemical combination, and the state in which the ingredient elements exist in steam, would have enabled any one to predict without a trial.

(38.) Ex. 3. (35.) II. After the invention of the diving-bell, and its success in subaqueous processes, it was considered highly desirable to devise some means of remaining for any length of time under water, and rising at pleasure without assistance, so as either to examine, at leisure, the bottom, or perform, at ease, any work that might be required. Some years ago, an ingenious individual proposed a project by which this end was to be accomplished. It consisted in sinking the hull of a ship made quite water-tight, with the decks and sides strongly supported by shores, and the only entry secured by a stout trap-door, in such a manner, that by disengaging, from within, the weights employed to sink it, it might rise of itself to the surface. To render the trial more satisfactory, and the result more striking, the projector himself made the first essay. It was agreed that he should sink in twenty fathoms water, and rise again without assistance at the expiration of twenty-four hours. Accordingly, making all secure, fastening down his trap-door, and provided with all necessaries, as well as with the means of making signals to indicate his situation, this unhappy victim of his own ingenuity entered and was sunk. No signal was made, and the time appointed elapsed. An immense concourse of people had assembled to witness his rising, but in vain; for the vessel was never seen more. The pressure of the water at so great a depth had, no doubt, been completely under-estimated, and the sides of the vessel being at once crushed in, the unfortunate projector perished before he could even make the signal concerted to indicate his distress.

(39.) Ex. 4. (35.) III. In the granite quarries near Seringapatam the most enormous blocks are separated from the solid rock by the following neat and simple process. The workman having found a portion of the rock sufficiently extensive, and situated near the edge of the part already quarried, lays bare the upper surface, and marks on it a line in the direction of the intended separation, along which a groove is cut with a chisel about a couple of inches in depth. Above this groove a narrow line of fire is then kindled, and maintained till the rock below is thoroughly heated, immediately on which a line of men and women, each provided with a pot full of cold water, suddenly sweep off the ashes, and pour the water into the heated groove, when the rock at once splits with a clean fracture. Square blocks of six feet in the side, and upwards of eighty feet in length, are sometimes detached by this method, or by another equally simple and efficacious, but not easily explained without entering into particulars of mineralogical detail.*

(40.) Ex. 5. (35.) III. Hardly less simple and efficacious is the process used in some parts of France, where mill-stones are made. When a mass of stone sufficiently large is found, it is cut into a cylinder several feet high, and the question then arises how to subdivide this into horizontal pieces so as to make as many mill-stones. For this purpose horizontal indentations or grooves are chiselled out quite round the cylinder, at distances corresponding to the thickness intended to be given to the mill-stones, into which wedges of dried wood are driven. These are then wetted, or exposed to the night dew, and next morning the different pieces are found separated from each other by the expansion of the wood, consequent on its absorption of moisture; an irresistible natural power thus accomplishing, almost without any trouble, and at no expense, an operation which, from the peculiar hardness and texture of the stone, would otherwise be impracticable but by the most powerful machinery or the most persevering labour.

(41.) Ex. 6. (35.) III. To accomplish our ends quickly is often of, at least, as much importance as to accomplish them with little labour and expense. There are innumerable processes which, if left to themselves, *i. e.* to the ordinary operation of natural causes, are done, and well done, but with extreme

* Such a block would weigh between four and five hundred thousand pounds. See Dr. Kennedy's "Account of the Erection of a Granite Obelisk of a Single Stone about Seventy Feet high, at Seringapatam."—*Ed. Phil. Trans.* vol. ix. p. 312.

slowness, and in such cases it is often of the highest practical importance to accelerate them. The bleaching of linen, for instance, performed in the natural way by exposure to sun, rain, and wind requires many weeks or even months for its completion; whereas, by the simple immersion of the cloth in a liquid, chemically prepared, the same effect is produced in a few hours. The whole circle of the arts, indeed, is nothing but one continued comment upon this head of our subject. The instances above given are selected, not on account of their superior importance, but for the simplicity and *directness* of application of the principles on which they depend, to the objects intended to be attained.

(42.) But so constituted is the mind of man, that his views enlarge, and his desires and wants increase, in the full proportion of the facilities afforded to their gratification, and, indeed, with augmented rapidity, so that no sooner has the successful exercise of his powers accomplished any considerable simplification or improvement of processes subservient to his use or comfort, than his faculties are again on the stretch to extend the limits of his newly acquired power; and having once experienced the advantages which are to be gathered by availing himself of some of the powers of nature to accomplish his ends, he is led thenceforward to regard them all as a treasure placed at his disposal, if he have only the art, the industry, or the good fortune, to penetrate those recesses which conceal them from immediate view. Having once learned to look on knowledge as power, and to avail himself of it as such, he is no longer content to limit his enterprises to the beaten track of former usage, but is constantly led onwards to contemplate objects which, in a previous stage of his progress, he would have regarded as unattainable and visionary, had he even thought of them at all. It is here that the investigation of the hidden powers of nature becomes a mine, every vein of which is pregnant with inexhaustible wealth, and whose ramifications appear to extend in all directions wherever human wants or curiosity may lead us to explore.

(43.) Between the physical sciences and the arts of life there subsists a constant mutual interchange of good offices, and no considerable progress can be made in the one without of necessity giving rise to corresponding steps in the other. On the one hand, every art is in some measure, and many entirely, dependent on those very powers and qualities of the material world which it is the object of physical enquiry to investigate and explain; and, accordingly, abundant examples might be cited of cases where the remarks of experienced artists, or even ordinary workmen, have led to the discovery of natural qualities, elements, or combinations which have proved of the highest importance in physics. Thus (to give an instance), a soap-manufacturer remarks that the residuum of his ley, when exhausted of the alkali for which he employs it, produces a corrosion of his copper boiler for which he cannot account. He puts it into the hands of a scientific chemist for analysis, and the result is the discovery of one of the most singular and important chemical elements, iodine. The properties of this, being studied, are found to occur most appositely in illustration and support of a variety of new, curious, and instructive views then gaining ground in chemistry, and thus exercise a marked influence over the whole body of that science. Curiosity is excited: the origin of the new substance is traced to the sea-plants from whose ashes the principal ingredient of soap is obtained, and ultimately to the sea-water itself. It is thence hunted through nature, discovered in salt mines and springs, and pursued into all bodies which have a marine origin; among the rest, into sponge. A medical practitioner * then calls to mind a reputed remedy for the cure of one of the most grievous and unsightly disorders to which the human species is subject — the *goître* — which infests the inhabitants of mountainous districts to an extent that in this favoured land we have happily no experience of, and which was said to have been originally cured by the ashes of burnt sponge. Led by this indication he tries the effect of iodine on that complaint, and the result establishes the extraordinary fact that this singular substance, taken as a medicine, acts with the utmost promptitude and energy on *goître*, dissipating the largest and most inveterate in a short time, and acting (of course, like all medicines, even the most approved, with occasional failures,) as a specific, or natural antagonist, against that odious deformity. It is thus that any accession to our knowledge of nature is sure, sooner or later, to make itself felt in some practical application, and that a benefit conferred on science by the casual observation or shrewd remark of even an unscientific or illiterate

* Dr. Coindet of Geneva.

person infallibly repays itself with interest, though often in a way that could never have been at first contemplated.

(44.) It is to such observation, reflected upon, however, and matured into a rational and scientific form by a mind deeply imbued with the best principles of sound philosophy, that we owe the practice of vaccination; a practice which has effectually subdued, in every country where it has been introduced, one of the most frightful scourges of the human race, and in some extirpated it altogether. Happily for us we know only by tradition the ravages of the small-pox, as it existed among us hardly more than a century ago, and as it would in a few years infallibly exist again, were the barriers which this practice, and that of inoculation, oppose to its progress abandoned. Hardly inferior to this terrible scourge on land was, within the last seventy or eighty years, the scurvy at sea. The sufferings and destruction produced by this horrid disorder on board our ships when, as a matter of course, it broke out after a few months' voyage, seem now almost incredible. Deaths to the amount of eight or ten a day in a moderate ship's company; bodies sewn up in hammocks and washing about the decks for want of strength and spirits on the part of the miserable survivors to cast them overboard; and every form of loathsome and excruciating misery of which the human frame is susceptible: — such are the pictures which the narratives of nautical adventure in those days continually offer.* At present the scurvy is almost

* Journal of a Voyage to the South Seas, &c. &c., under the Command of Commodore George Anson in 1740—1744, by

completely eradicated in the navy, partly, no doubt, from increased and increasing attention to general cleanliness, comfort, and diet; but mainly from the constant use of a simple and palatable preventive, the acid of the lemon, served out in daily rations. If the gratitude of mankind be allowed on all hands to be the just meed of the philosophic physician, to whose discernment in seizing, and perseverance in forcing it on public notice we owe the great safeguard of infant life, it ought not to be denied to those * whose skill and discrimination have thus

---

Pascoe Thomas, Lond. 1745. So tremendous were the ravages of scurvy, that, in the year 1726, admiral Hosier sailed with seven ships of the line to the West Indies, and buried his ships' companies twice, and died himself in consequence of a broken heart. Dr. Johnson, in the year 1778, could describe a sea-life in such terms as these: —" As to the sailor, when you look down from the quarter deck to the space below, you see the utmost extremity of human misery, such crowding, such filth, such stench!"—" A ship is a prison with the chance of being drowned — it is worse — worse in every respect — worse room, worse air, worse food — worse company!" Smollet, who had personal experience of the horrors of a seafaring life in those days, gives a lively picture of them in his Roderick Random.

* Lemon juice was known to be a remedy for scurvy far superior to all others 200 years ago, as appears by the writings of Woodall. His work is entitled " The Surgeon's Mate, or Military and Domestic Medicine. By John Woodall, Master in Surgery London, 1636," p. 165. In 1600, Commodore Lancaster sailed from England with three other ships for the Cape of Good Hope, on the 2d of April, and arrived in Saldanha Bay on the 1st of August, the commodore's own ship being in perfect health, from the administration of three table-spoonsfull of lemon juice every morning to each of his men, whereas the other ships were so sickly as to be un-

strengthened the sinews of our most powerful arm, and obliterated one of the darkest features in the most glorious of all professions.

(45.) These last, however, are instances of simple observation, limited to the point immediately in view, and assuming only so far the character of science as a systematic adoption of good and rejection of evil, when grounded on experience carefully weighed, justly entitle it to do. They are not on that account less appositely cited as instances of the importance

---

manageable for want of hands, and the commander was obliged to send men on board to take in their sails and hoist out their boats. (Purchas's Pilgrim, vol. i. p. 149.) A Fellow of the college, and an eminent practitioner, in 1753 published a tract on sea scurvy, in which he adverts to the superior virtue of this medicine; and Mr. A. Baird, surgeon of the Hector sloop of war, states, that from what he had seen of its effects on board of that ship, he "thinks he shall not be accused of presumption in pronouncing it, if properly administered, a *most infallible remedy*, both in the cure and prevention of scurvy." (Vide Trotter's Medicina Nautica.) The precautions adopted by captain Cook in his celebrated voyages, had fully demonstrated by their complete success the practicability of keeping scurvy under in the longest voyages, but a uniform system of prevention throughout the service was still deficient.

It is to the representations of Dr. Blair and sir Gilbert Blane, in their capacity of commissioners of the board for sick and wounded seamen, in 1795, we believe, that its *systematic introduction into nautical diet*, by a general order of the admiralty, is owing. The effect of this wise measure (taken, of course, in conjunction with the general causes of improved health,) may be estimated from the following facts: — In 1780, the number of cases of scurvy received into Haslar hospital was 1457; in 1806 *one* only, and in 1807 *one*. There are now many surgeons in the navy who have never seen the disease.

of a knowlege of nature and its laws to our well-being; though, like the great inventions of the mariner's compass and of gunpowder, they may have stood, in their origin, unconnected with more general views. They are rather to be looked upon as the spontaneous produce of a territory essentially fertile, than as forming part of the succession of harvests which the same bountiful soil, diligently cultivated, is capable of yielding. The history of iodine above related affords, however, a perfect specimen of the manner in which a knowledge of natural properties and laws, collected from facts having no reference to the object to which they have been subsequently applied, enables us to set in array the resources of nature against herself; and deliberately, of afore-thought, to devise remedies against the dangers and inconveniences which beset us. In this view we might instance, too, the *conductor*, which, in countries where thunder-storms are more frequent and violent than in our own, and at sea (where they are attended with peculiar danger, both from the greater probability of accident, and its more terrible consequences when it does occur,) forms a most real and efficient preservative against the effects of lightning * :— the *safety-lamp*, which enables us to walk with light and

* Throughout France the conductor is recognised as a most valuable and useful instrument; and in those parts of Germany where thunder-storms are still more common and tremendous they are become nearly universal. In Munich there is hardly a modern house unprovided with them, and of a much better construction than ours — several copper wires twisted into a rope.

security while surrounded with an atmosphere more explosive than gunpowder: — the *life-boat*, which cannot be sunk, and which offers relief in circumstances of all others the most distressing to humanity, and of which a recent invention promises to extend the principle to ships of the largest class: — the *lighthouse*, with the capital improvements which the lenses of Brewster and Fresnel, and the elegant lamp of lieutenant Drummond, have conferred, and promise yet to confer by their wonderful powers, the one of producing the most intense light yet known, the others of conveying it undispersed to great distances: — the discovery of the disinfectant powers of chlorine, and its application to the destruction of miasma and contagion: — that of *quinine*, the essential principle in which reside the febrifuge qualities of the Peruvian bark, a discovery by which posterity is yet to benefit in its full extent, but which has already begun to diffuse *comparative* comfort and health through regions almost desolated by pestiferous exhalations *; — and, if we desist, it is not because the list is exhausted, but because a sample, not a catalogue, is intended.

(46.) One instance more, however, we will add, to illustrate the manner in which a most familiar effect, which seemed destined only to amuse children, or, at best, to furnish a philosophic toy, may become a safeguard of human life, and a remedy for a most

* We have been informed by an eminent physician in Rome, (Dr. Morichini) that a vast quantity of the sulphate of quinine is manufactured there and consumed in the Campagna, with an evident effect in mitigating the severity of the malarious complaints which affect its inhabitants.

serious and distressing evil. In needle manufactories the workmen who point the needles are constantly exposed to excessively minute particles of steel which fly from the grindstones, and mix, though imperceptible to the eye, as the finest dust in the air, and are inhaled with their breath. The effect, though imperceptible on a short exposure, yet, being constantly repeated from day to day, produces a constitutional irritation dependent on the tonic properties of the steel, which is sure to terminate in pulmonary consumption; insomuch, that persons employed in this kind of work used scarcely ever to attain the age of forty years.* In vain was it attempted to purify the air before its entry into the lungs by gauzes or linen guards; the dust was too fine and penetrating to be obstructed by such coarse expedients, till some ingenious person bethought him of that wonderful power which every child who searches for its mother's needle with a magnet, or admires the motions and arrangement of a few steel filings on a sheet of paper held above it, sees in exercise. Masks of magnetized steel wire are now constructed and adapted to the faces of the workmen. By these the air is not merely *strained* but *searched* in its passage through them, and each obnoxious atom arrested and removed.

(47.) Perhaps there is no result which places in a stronger light the advantages which are to be derived from a mere knowledge of the *usual order of nature,* without any attempt on our part to modify it, and apart from all consideration of its causes,

* Dr. Johnson, Memoirs of the Medical Society, vol. v.

than the institution of life-assurances. Nothing is more uncertain than the life of a single individual; and it is the sense of this insecurity which has given rise to such institutions. They are, in their nature and objects, the precise reverse of gambling speculations, their object being to equalize vicissitude, and to place the pecuniary relations of numerous masses of mankind, in so far as they extend, on a footing independent of individual casualty. To do this with the greatest possible advantage, or indeed with any advantage at all, it is necessary to know the *laws of mortality*, or the average numbers of individuals, out of a great multitude, who die at every period of life from infancy to extreme old age. At first sight this would seem a hopeless enquiry; to some, perhaps, a presumptuous one. But it has been made; and the result is, that, abating extraordinary causes, such as wars, pestilence, and the like, a remarkable regularity *does* obtain, quite sufficient to afford grounds not only for general estimations, but for nice calculations of risk and adventure, such as infallibly to insure the success of any such institution founded on good computations; and thus to confer such stability on the fortunes of families dependent on the exertions of one individual as to constitute an important feature in modern civilization. The only thing to be feared in such institutions is their too great multiplication and consequent competition, by which a spirit of gambling and underbidding is liable to be generated among their conductors, and the very mischief may be produced, on a scale of frightful extent, which they are especially intended to prevent.

(48.) We have hitherto considered only cases in which a knowledge of natural laws enables us to improve our condition, by counteracting evils of which, but for its possession, we must have remained for ever the helpless victims. Let us now take a similar view of those in which we are enabled to call in nature as an auxiliary to augment our actual power, and capacitate us for undertakings, which without such aid might seem to be hopeless. Now, to this end, it is necessary that we should form a just conception of what those hidden powers of nature *are*, which we can at pleasure call into action; — how far they transcend the measure of human force, and set at naught the efforts not only of individuals but of whole nations of men.

(49.) It is well known to modern engineers, that *there is virtue* in a bushel of coals properly consumed, to raise seventy millions of pounds weight a foot high. This is actually the *average* effect of an engine at this moment working in Cornwall.* Let us pause a moment, and consider what this is equivalent to in matters of practice.

(50.) The ascent of Mont Blanc from the valley of Chamouni is considered, and with justice, as the most toilsome feat that a strong man can execute in two days. The combustion of two pounds of coal would place him on the summit.†

* The engine at Huel Towan. See Mr. Henwood's Statement "of the performance of steam-engines in Cornwall for April, May, and June, 1829." Brewster's Journal, Oct. 1829. —The *highest* monthly average of this engine extends to 79 millions of pounds.

† However, this is not quite a fair statement; a man's daily

(51.) The Menai Bridge, one of the most stupendous works of art that has been raised by man in modern ages, consists of a mass of iron, not less than four millions of pounds in weight, suspended at a medium height of about 120 feet above the sea. The consumption of seven bushels of coal would suffice to raise it to the place where it hangs.

(52.) The great pyramid of Egypt is composed of granite. It is 700 feet in the side of its base, and 500 in perpendicular height, and stands on eleven acres of ground. Its weight is, therefore, 12,760 millions of pounds, at a medium height of 125 feet; consequently it would be raised by the effort of about 630 chaldrons of coal, a quantity consumed in some founderies in a week.

(53.) The annual consumption of coal in London is estimated at 1,500,000 chaldrons. The effort of this quantity would suffice to raise a cubical block of marble, 2200 feet in the side, through a space equal to its own height, or to pile one such mountain upon another. The Monte Nuovo, near Pozzuoli, (which was erupted in a single night by volcanic fire,) might have been raised by such an effort, from a depth of 40,000 feet, or about eight miles.

(54.) It will be observed, that, in the above statement, the inherent power of fuel is, of necessity, greatly under-rated. It is not pretended by engineers that the economy of fuel is yet pushed to its utmost limit, or that the whole effective power is obtained in any application of fire yet devised; so that were

---

labour is about 4 lbs. of coals. The extreme toil of this ascent arises from other obvious causes than the mere height.

we to say 100 millions instead of 70, we should probably be nearer the truth.

(55.) The powers of wind and water, which we are constantly impressing into our service, can scarcely be called latent or hidden, yet it is not fully considered, in general, what they *do* effect for us. Those who would judge of what advantage may be taken of the wind, for example, even on land (not to speak of navigation), may turn their eyes on Holland. A great portion of the most valuable and populous tract of this country lies much below the level of the sea, and is only preserved from inundation by the maintenance of embankments. Though these suffice to keep out the abrupt influx of the ocean, they cannot oppose that law of nature, by which fluids, in seeking their level, insinuate themselves through the pores and subterraneous channels of a loose sandy soil, and keep the country in a constant state of infiltration from below upwards. To counteract this tendency, as well as to get rid of the rain water, which has no natural outlet, pumps worked by windmills are established in great numbers, on the dams and embankments, which pour out the water, as from a leaky ship, and in effect preserve the country from submersion, by taking advantage of every wind that blows. To drain the Haarlem lake * would seem a hopeless project to any speculators but those who had the steam-engine at their command, or had learnt in

* Its surface is about 40,000 acres, and medium depth about 20 feet. It was proposed to drain it by running embankments across it, and thus cutting it up into more manageable portions to be drained by windmills.

Holland what might be accomplished by the constant agency of the desultory but unwearied powers of wind. But the Dutch engineer measures his surface, calculates the number of his pumps, and, trusting to time and his experience of the operation of the winds for the success of his undertaking, boldly forms his plans to lay dry the bed of an inland sea, of which those who stand on one shore cannot see the other.*

(56.) To gunpowder, as a source of mechanical power, it seems hardly necessary to call attention; yet it is only when we endeavour to *confine* it, that we get a full conception of the immense energy of that astonishing agent. In count Rumford's experiments, twenty-eight grains of powder confined in a cylindrical space, *which it just filled*, tore asunder a piece of iron which would have resisted a strain of 400,000 lbs.†, applied at no greater mechanical disadvantage.

(57.) But chemistry furnishes us with means of calling into sudden action forces of a character infinitely more tremendous than that of gunpowder. The terrific violence of the different fulminating compositions is such, that they can only be compared to those untameable animals, whose ferocious

* No one doubts the *practicability* of the undertaking. Eight or nine thousand chaldrons of coals duly burnt would evacuate the whole contents. But many doubt whether it would be profitable, and some, considering that a few hundreds of fishermen who gain their livelihood on its waters would be dispossessed, deny that it would be *desirable*.

† "Experiments to determine the Force of fired Gunpowder." Phil. Trans. vol. lxxxvii. p. 254. et seq.

strength has hitherto defied all useful management, or rather to spirits evoked by the spells of a magician, manifesting a destructive and unapproachable power, which makes him but too happy to close his book, and break his wand, as the price of escaping unhurt from the storm he has raised. Such powers are not yet subdued to our purposes, whatever they may hereafter be; but, in the expansive force of gases, liberated slowly and manageably from chemical mixtures, we have a host of inferior, yet still most powerful, energies, capable of being employed in a variety of useful ways, according to emergencies.*

(58.) Such are the forces which nature lends us for the accomplishment of our purposes, and which it is the province of practical Mechanics to teach us to combine and apply in the most advantageous manner; without which the mere command of power would amount to nothing. Practical Mechanics is, in the most pre-eminent sense, a *scientific art;* and it may be truly asserted, that almost all the great combinations of modern mechanism, and many of its refinements and nicer improvements, are creations of pure intellect, grounding its exertion upon a moderate number of very elementary propositions in theoretical mechanics and geometry. On this head we might dwell long, and find ample matter, both

* See a very ingenious application of this kind in Mr. Babbage's article on Diving in the Encyc. Metrop. — Others will readily suggest themselves. For instance, the ballast in reserve of a balloon might consist of materials capable of evolving great quantities of hydrogen gas in proportion to their weight, should such be found.

for reflection and wonder; but it would require not volumes merely, but libraries, to enumerate and describe the prodigies of ingenuity which have been lavished on every thing connected with machinery and engineering. By these it is that we are enabled to diffuse over the whole earth the productions of any part of it; to fill every corner of it with miracles of art and labour, in exchange for its peculiar commodities; and to concentrate around us, in our dwellings, apparel and utensils, the skill and workmanship not of a few expert individuals, but of all who, in the present and past generations, have contributed their improvements to the processes of our manufactures.

(59.) The transformations of chemistry, by which we are enabled to convert the most apparently useless materials into important objects in the arts, are opening up to us every day sources of wealth and convenience of which former ages had no idea, and which have been pure gifts of science to man. Every department of art has felt their influence, and new instances are continually starting forth of the unlimited resources which this wonderful science developes in the most sterile parts of nature. Not to mention the impulse which its progress has given to a host of other sciences, which will come more particularly under consideration in another part of this discourse, what strange and unexpected results has it not brought to light in its application to some of the most common objects! Who, for instance, would have conceived that linen rags were capable of producing *more than their own weight* of sugar, by the simple agency of one of the cheapest and most abundant acids? *—that dry bones could be a magazine of nutriment, capable of preservation for years, and ready to yield up their sustenance in the form best adapted to the support of life, on the application of that powerful agent, steam, which enters so largely into all our processes, or of an acid at once cheap and durable? †— that sawdust itself is susceptible of conversion into a substance bearing no remote analogy to bread; and though certainly less palatable than that of flour, yet no way disagreeable, and both wholesome and digestible as well as highly nutritive? ‡ What economy, in all processes where chemical agents are employed, is introduced by the exact knowledge of the proportions in which natural elements unite, and their mutual powers of displacing each other! What perfection in all the arts where fire is employed, either in its more violent applications, (as, for instance, in the smelting of metals by the introduction of well adapted fluxes, whereby we obtain the whole produce of the ore in its purest state,) or in its milder forms, as in sugar-refining (the whole modern practice of which depends on a curious and delicate remark of a late eminent scientific chemist on the nice adjustment of temperature at which the crystallization of syrup takes place); and a thousand other arts which it would be tedious to enumerate!

* The sulphuric. Bracconot, Annales de Chimie, vol. xii. p. 184.

† D'Arcet, Annales de l'Industrie, Février, 1829.

‡ See Dr. Prout's account of the experiments of professor Autenrieth of Tubingen. Phil. Trans. 1827. p. 381. This discovery, which renders famine next to *impossible*, deserves a higher degree of celebrity than it has obtained.

(60.) Armed with such powers and resources, it is no wonder if the enterprise of man should lead him to form and execute projects which, to one uninformed of their grounds, would seem altogether disproportionate. Were they to have been proposed at once, we should, no doubt, have rejected them as such: but developed, as they have been, in the slow succession of ages, they have only taught us that things regarded impossible in one generation may become easy in the next; and that the power of man over nature is limited only by the one condition, that it must be exercised in comformity with the laws of nature. He must study those laws as he would the disposition of a horse he would ride, or the character of a nation he would govern; and the moment he presumes either to thwart her fundamental rules, or ventures to measure his strength with hers, he is at once rendered severely sensible of his imbecility, and meets the deserved punishment of his rashness and folly. But if, on the other hand, he will consent to use, without abusing, the resources thus abundantly placed at his disposal, and obey that he may command, there seems scarcely any conceivable limit to the degree in which the *average* physical condition of great masses of mankind may be improved, their wants supplied, and their conveniences and comforts increased. Without adopting such an exaggerated view, as to assert that the meanest inhabitant of a civilized society is superior in physical condition to the lordly savage, whose energy and uncultivated ability gives him a natural predominance over his fellow denizens of the forest,—at least, if we compare like with like, and consider the multitude of human beings who are enabled, in an advanced state of society, to subsist in a degree of comfort and abundance, which at best only a few of the most fortunate in a less civilized state could command, we shall not be at a loss to perceive the principle on which we ought to rest our estimate of the advantages of civilization; and which applies with hardly less force to every degree of it, when contrasted with that next inferior, than to the broad distinction between civilized and barbarous life in general.

(61.) The difference of the degrees in which the individuals of a great community enjoy the good things of life has been a theme of declamation and discontent in all ages; and it is doubtless our paramount duty, in every state of society, to alleviate the pressure of the purely evil part of this distribution as much as possible, and, by all the means we can devise, secure the lower links in the chain of society from dragging in dishonour and wretchedness: but there is a point of view in which the picture is at least materially altered in its expression. In comparing society on its present immense scale, with its infant or less developed state, we must at least take care to enlarge every feature in the same proportion. If, on comparing the *very* lowest states in civilized and savage life, we admit a difficulty in deciding to which the preference is due, at least in every superior grade we cannot hesitate a moment; and if we institute a similar comparison in every different stage of its progress, we cannot fail to be struck with the rapid *rate of dilatation* which every degree upward of the scale, so to speak, exhibits, and which, in an estimate of averages, gives an immense preponderance to the present over every former condition of mankind, and, for aught we can see to the contrary, will place succeeding generations in the same degree of superior relation to the present that this holds to those passed away. Or we may put the same proposition in other words, and, admitting the existence of every inferior grade of advantage in a higher state of civilization which subsisted in the preceding, we shall find, first, that, taking state for state, the proportional numbers of those who enjoy the higher degrees of advantage increases with a constantly accelerated rapidity as society advances; and, secondly, that the superior extremity of the scale is constantly enlarging by the addition of new degrees. The condition of a European prince is now as far superior, in the command of real comforts and conveniences, to that of one in the middle ages, as that to the condition of one of his own dependants.

(62.) The advantages conferred by the augmentation of our physical resources through the medium of increased knowledge and improved art have this peculiar and remarkable property, — that they are in their nature diffusive, and cannot be enjoyed in any exclusive manner by a few. An eastern despot may extort the riches and monopolize the art of his subjects for his own personal use; he may spread around him an unnatural splendour and luxury, and stand in strange and preposterous contrast with the general penury and discomfort of his people; he may glitter in jewels of gold and raiment of needlework; but the wonders of well contrived and executed manufacture which we use daily, and the comforts which have been invented, tried, and improved upon by thousands, in every form of domestic convenience, and for every ordinary purpose of life, can never be enjoyed by him. To produce a state of things in which the physical advantages of civilized life can exist in a high degree, the stimulus of increasing comforts and constantly elevated desires, must have been felt by millions; since it is not in the power of a few individuals to create that wide demand for useful and ingenious applications, which alone can lead to great and rapid improvements, unless backed by that arising from the speedy diffusion of the same advantages among the mass of mankind.

(63.) If this be true of physical advantages, it applies with still greater force to intellectual. Knowledge can neither be adequately cultivated nor adequately enjoyed by a few; and although the conditions of our existence on earth may be such as to preclude an abundant supply of the physical necessities of all who may be born, there is no such law of nature in force against that of our intellectual and moral wants. Knowledge is not, like food, destroyed by use, but rather augmented and perfected. It acquires not, perhaps, a greater certainty, but at least a confirmed authority and a probable duration, by universal assent; and there is no body of knowledge so complete, but that it may acquire accession, or so free from error but that it may receive correction in passing through the minds of millions. Those who admire and love knowledge for its own sake ought to wish to see its elements made accessible to all, were it only that they may be the more thoroughly examined into, and more effectually developed in their consequences, and receive that ductility and plastic quality which the pressure of minds of all descriptions, constantly moulding them to their purposes, can alone bestow. But to this end it is necessary that it should be divested, as far as possible, of artificial difficulties, and stripped of all such technicalities as tend to place it in the light of a craft and a mystery, inaccessible without a kind of apprenticeship. Science, of course, like every thing else, has its own peculiar terms, and, so to speak, its idioms of language; and these it would be unwise, were it even possible, to relinquish: but every thing that tends to clothe it in a strange and repulsive garb, and especially every thing that, to keep up an appearance of superiority in its professors over the rest of mankind, assumes an unnecessary guise of profundity and obscurity, should be sacrificed without mercy. Not to do this, is to deliberately reject the light which the natural unencumbered good sense of mankind is capable of throwing on every subject, even in the elucidation of principles: but where principles are to be applied to practical uses it becomes absolutely necessary; as all mankind have then an interest in their being so familiarly understood, that no mistakes shall arise in their application.

(64.) The same remark applies to arts. They cannot be perfected till their whole processes are laid open, and their language simplified and rendered universally intelligible. Art is the application of knowledge to a practical end. If the knowledge be merely accumulated experience, the art is *empirical;* but if it be experience reasoned upon and brought under general principles, it assumes a higher character, and becomes a *scientific. art.* In the progress of mankind from barbarism to civilised life, the arts necessarily precede science. The wants and cravings of our animal constitution must be satisfied; the comforts, and some of the luxuries, of life must exist. Something must be given to the vanity or show, and more to the pride of power: the round of baser pleasures must have been tried and found insufficient, before intellectual ones can gain a footing; and when they have obtained it, the delights of poetry and its sister arts still take precedence of contemplative enjoyments, and the severer pursuits of thought; and when these in time begin to charm from their novelty, and sciences begin to arise, they will at first be those of pure speculation. The mind delights to escape from the trammels which had bound it to earth, and luxuriates in its newly found powers. Hence, the abstractions of geometry — the properties of numbers — the movements of the celestial spheres — whatever is abstruse, remote, and extramundane — become the first objects of infant science. Applications come late: the arts continue slowly progressive, but their realm remains separated from that of science by a wide gulf which can only be passed by a powerful spring. They form their own language and their own conventions, which none but artists can understand. The whole tendency of empirical art, is to bury itself in technicalities, and to place its pride in particular short cuts and mysteries known only to adepts; to surprise and astonish by results, but conceal processes. The character of science is the direct contrary. It delights to lay itself open to enquiry; and is not satisfied with its conclusions, till it can make the road to them broad and beaten: and in its applications it preserves the same character; its whole aim being to strip away all technical mystery, to illuminate every dark recess, and to gain free access to all processes, with a view to improve them on rational principles. It would seem that a union of two qualities almost opposite to each other — a going forth of the thoughts in two directions, and a sudden transfer of ideas from a remote station in one to an equally distant one in the other—is required to start the first idea of *applying science.* Among the Greeks, this point was attained by Archimedes, but attained too late, on the eve of that great eclipse of science which was destined to continue for nearly eighteen centuries, till Galileo in Italy, and Bacon in England, at once dispelled the darkness: the one, by his inventions and discoveries; the other, by the irresistible force of his arguments and eloquence.

(65.) Finally, the improvement effected in the condition of mankind by advances in physical science as applied to the useful purposes of life, is very far from being limited to their direct consequences in the more abundant supply of our physical wants, and the increase of our comforts. Great as these benefits are, they are yet but steps to others of a still higher kind. The successful results of our experiments and reasonings in natural philosophy, and the incalculable advantages which experience, systematically consulted and dispassionately reasoned on, has conferred in matters purely physical, tend of necessity to impress something of the well weighed and progressive character of science on the more complicated conduct of our social and moral relations. It is thus that legislation and politics become gradually regarded as experimental sciences; and history, not, as formerly, the mere record of tyrannies and slaughters, which, by immortalizing the execrable actions of one age, perpetuates the ambition of committing them in every succeeding one, but as the archive of experiments, successful and unsuccessful, gradually accumulating towards the solution of the grand problem — how the advantages of government are to be secured with the least possible inconvenience to the governed. The celebrated apophthegm, that nations never profit by experience, becomes yearly more and more untrue. Political economy, at least, is found to have sound principles, founded in the moral and physical nature of man, which, however lost sight of in particular measures — however even temporarily controverted and borne down by clamour — have yet a stronger and stronger testimony borne to them in each succeeding generation, by which they must, sooner or later, prevail. The idea once conceived and verified, that great and noble ends are to be achieved, by which the condition of the whole human species shall be permanently bettered, by bringing into exercise a sufficient quantity of sober thought, and by a proper adaptation of means, is of itself sufficient to set us earnestly on reflecting what ends *are* truly great and noble, either in themselves, or as conducive to others of a still loftier character; because we are not now, as heretofore, hopeless of attaining them. It is not now equally harmless and insignificant, whether we are right or wrong; since we are no longer supinely and helplessly carried down the stream of events, but feel ourselves capable of buffetting at least with its waves, and perhaps of riding triumphantly over them: for why should we despair that the reason which has enabled us to subdue all nature to our purposes, should (if permitted and assisted by the providence of God) achieve a far more difficult conquest; and ultimately find some means of enabling the collective wisdom of mankind to bear down those obstacles which individual short-sightedness, selfishness, and passion, oppose to all improvements, and by which the highest hopes are continually blighted, and the fairest prospects marred.

PART II.

OF THE PRINCIPLES ON WHICH PHYSICAL SCIENCE RELIES FOR ITS SUCCESSFUL PROSECUTION, AND THE RULES BY WHICH A SYSTEMATIC EXAMINATION OF NATURE SHOULD BE CONDUCTED, WITH ILLUSTRATIONS OF THEIR INFLUENCE AS EXEMPLIFIED IN THE HISTORY OF ITS PROGRESS.

CHAPTER I.

OF EXPERIENCE AS THE SOURCE OF OUR KNOWLEDGE. — OF THE DISMISSAL OF PREJUDICES. — OF THE EVIDENCE OF OUR SENSES.

(66.) INTO abstract science, as we have before observed, the notion of cause does not enter. The truths it is conversant with are *necessary* ones, and exist independent of cause. There may be no such real *thing* as a right-lined triangle marked out in space; but the moment we conceive one in our minds, we cannot refuse to admit the sum of its three angles to be equal to two right angles; and if in addition we conceive one of its angles to be a right angle, we cannot thenceforth refuse to admit that the sum of the squares on the two sides, including the right angle, is equal to the square on the side subtending it. To maintain the contrary, would be, in effect, to deny its being right angled. No one *causes* or *makes* all the diameters of an ellipse to be bisected in its centre. To assert the contrary, would not be to rebel against a power, but to deny our own words. But in natural science *cause* and *effect* are the ultimate relations we contemplate; and *laws*, whether imposed or maintained, which, for aught we can perceive, might have been other than they are. This distinction is very important. A clever man, shut up alone and allowed unlimited time, might reason out for himself all the truths of mathematics, by proceeding from those simple notions of space and number of which he cannot divest himself without ceasing to think. But he could never tell, by any effort of reasoning, what would become of a lump of sugar if immersed in water, or what impression would be produced on his eye by mixing the colours yellow and blue.

(67.) We have thus pointed out to us, as the great, and indeed only ultimate source of our knowledge of nature and its laws, EXPERIENCE; by which we mean, not the experience of one man only, or of one generation, but the accumulated experience of all mankind in all ages, registered in books or recorded by tradition. But experience may be acquired in two ways: either, first, by noticing facts as they occur, without any attempt to influence the frequency of their occurrence, or to vary the circumstances under which they occur; this is OBSERVATION: or, secondly, by putting in action causes and agents over which we have control, and purposely varying their combinations, and noticing what effects take place; this is EXPERIMENT. To these two sources we must look as the fountains of all natural science. It is not intended, however, by thus distinguishing observation from experiment, to place them in any kind of contrast. Essentially they are much alike, and differ rather in degree than in kind; so that, perhaps, the terms *passive* and *active observation* might better express their distinction; but it is, nevertheless, highly important to mark the different states of mind in inquiries carried on by their respective aids, as well as their different effects in promoting the progress of science. In the former, we sit still and listen to a tale, told us, perhaps obscurely, piecemeal, and at long intervals of time, with our attention more or less awake. It is only by after-rumination that we gather its full import; and often, when the opportunity is gone by, we have to regret that our attention was not more particularly directed to some point which, at the time, appeared of little moment, but of which we at length appretiate the importance. In the latter, on the other hand, we cross-examine our witness, and by comparing one part of his evidence with the other, while he is yet before us, and reasoning upon it in his presence, are enabled to put pointed and searching questions, the answer to which may at once enable us to make up our minds. Accordingly it has been found invariably, that in those departments of physics where the phenomena are beyond our control, or into which experimental enquiry, from other causes, has not been carried, the progress of knowledge has been slow, uncertain, and irregular; while in such as admit of experiment, and in which mankind have agreed to its adoption, it has been rapid, sure, and steady. For example, in our knowledge of the nature and causes of volcanoes, earthquakes, the fall of stones from the sky, the appearance of new stars and disappearance of old ones, and other of those great phenomena of nature which are altogether beyond our command, and at the same time are of too rare occurrence to permit any one to repeat and rectify his impressions respecting them, we know little more now than in the earliest times. Here our tale is told us slowly, and in broken sentences. In astronomy, again, we have at least an uninterrupted narrative; the opportunity of observation is constantly present, and makes up in some measure for the impossibility of varying our point of view, and calling for information at the precise moment it is wanted. Accordingly, astronomy, regarded as a science of mere observation, arrived, though by very slow degrees, to a state of considerable maturity. But the moment that it became a branch of mechanics, a science essentially experimental, (that is to say, one in which any principle laid down can be subjected to immediate and decisive *trial*, and where experience does not require to be waited for,) its progress suddenly acquired a tenfold acceleration; nay, to such a degree, that it has been asserted, and we believe with truth, that were the records of all observations from the earliest ages annihilated, leaving only those made in a single observatory*, during a single lifetime†, the whole of this most perfect of sciences might, from those data, and as to the objects included in them, be at once recon-

* Greenwich. † Maskelyne's.

structed, and appear precisely as it stood at their conclusion. To take another instance: mineralogy, till modern times, could hardly be said to exist. The description of even the precious stones in Theophrastus and Pliny are, in most cases, hardly sufficient to identify them, and in many fall short even of that humble object; more recent observers, by attending more carefully to the obvious characters of minerals, had formed a pretty extensive catalogue of them, and made various attempts to arrange and methodize the knowledge thus acquired, and even to deduce some general conclusions respecting the forms they habitually assume: but from the moment that chemical analysis was applied to resolve them into their constituent elements, and that, led by a happy accident, the genius of Bergmann discovered the general fact, that they could be *cloven* or split in such directions as to lay bare their peculiar primitive or fundamental forms, (which lay concealed within them, as the statue might be conceived encrusted in its marble envelope,) — from that moment, mineralogy ceased to be an unmeaning list of names, a mere laborious cataloguing of stones and rubbish, and became, what it now is, a regular, methodical, and most important science, in which every year is bringing to light new relations, new laws, and new practical applications.

(68.) Experience once recognized as the fountain of all our knowledge of nature, it follows that, in the study of nature and its laws, we ought at once to make up our minds to dismiss as idle prejudices, or at least suspend as premature, any preconceived notion of what might or what ought to be the order of nature in any proposed case, and content ourselves with observing, as a plain matter of fact, what *is*. To experience we refer, as the only ground of all physical enquiry. But before experience itself can be used with advantage, there is one preliminary step to make, which depends wholly on ourselves: it is the absolute dismissal and clearing the mind of all prejudice, from whatever source arising, and the determination to stand and fall by the result of a direct appeal to facts in the first instance, and of strict logical deduction from them afterwards. Now, it is necessary to distinguish between two kinds of prejudices, which exercise very different dominion over the mind, and, moreover, differ extremely in the difficulty of dispossessing them, and the process to be gone through for that purpose. These are, —

1. Prejudices of opinion.
2. Prejudices of sense.

(69.) By prejudices of opinion, we mean opinions hastily taken up, either from the assertion of others, from our own superficial views, or from vulgar observation, and which, from being constantly admitted without dispute, have obtained the strong hold of habit on our minds. Such were the opinions once maintained that the earth is the greatest body in the universe, and placed immovable in its centre, and all the rest of the universe created for its sole use; that it is the nature of fire and of sounds to ascend; that the moonlight is cold; that dews *fall* from the air, &c.

(70.) To combat and destroy such prejudices we may proceed in two ways, either by demonstrating the falsehood of the facts alleged in their support, or by showing how the appearances, which seem to countenance them, are more satisfactorily accounted for without their admission. But it is unfortunately the nature of prejudices of opinion to adhere, in a certain degree, to every mind, and to some with pertinacious obstinacy, *pigris radicibus*, after all ground for their reasonable entertainment is destroyed. Against such a disposition the student of natural science must contend with all his power. Not that we are so unreasonable as to demand of him an instant and peremptory dismission of all his former opinions and judgments; all we require is, that he will hold them without bigotry, retain till he shall see reason to question them, and be ready to resign them when fairly proved untenable, and to doubt them when the weight of probability is shown to lie against them. If he refuse this, he is incapable of science.

(71.) Our resistance against the destruction of the other class of prejudices, those of sense, is commonly more violent at first, but less persistent, than in the case of those of opinion. Not to trust the evidence of our senses, seems, indeed, a hard condition, and one which, if proposed, none would comply with. But it is not the direct evidence of our senses that we are in any case called upon to reject, but only the erroneous judgments we unconsciously form from them, and this only when they can be shown to be so *by counter evidence of the same sort;* when one sense is brought to testify against another, for instance; or the same sense against itself, and the obvious conclusions in the two cases disagree, so as to compel us to acknowledge that one or other must be wrong. For example, nothing at first can seem a more rational, obvious, and incontrovertible conclusion, than that the *colour* of an object is an inherent quality, like its weight, hardness, &c. and that to *see* the object, and see it *of its own colour*, when nothing intervenes between our eyes and it, are one and the same thing. Yet this is only a prejudice; and that it is so, is shown by bringing forward the same sense of vision which led to its adoption, as evidence on the other side; for, when the differently coloured prismatic rays are thrown, in a dark room, in succession upon any object, whatever be the colour we are in the habit of calling its own, it will appear of the particular hue of the light which falls upon it: a yellow paper, for instance, will appear scarlet when illuminated by red rays, yellow when by yellow, green by green, and blue by blue rays; its own (so called) proper colour *not in the least degree mixing with that it so exhibits.*

(72.) To give one or two more examples of the kind of illusion which the senses practise on us, or rather which we practise on ourselves, by a misinterpretation of their evidence: the moon at its rising and setting appears much larger than when high up in the sky. This is, however, a mere erroneous judgment; for when we come to measure its diameter, so far from finding our conclusion borne out by fact, we actually find it to measure materially less. Here is eyesight opposed to eyesight, with the advantage of deliberate measurement. In ventriloquism we have the hearing at variance with all the other senses, and especially with the sight, which is sometimes contradicted by it in a very extraordinary and surprising manner, as when the voice is made to seem to issue from an inanimate and motionless object. If we plunge our hands, one into ice-cold water, and the other into water as hot as can be borne, and, after letting them stay awhile, suddenly transfer them both to a vessel full of water at a blood heat, the one will feel a sensation of heat, the other of cold. And if we cross the two first fingers of one hand, and place a pea in the fork between them, moving and rolling it about on a table, we shall (especially if we close our eyes) be fully persuaded we have two peas. If the nose be held while we are eating cinnamon, we shall perceive no difference between its flavour and that of a deal shaving.

(73.) These, and innumerable instances we might cite, will convince us, that though we are never deceived in the *sensible impression* made by external objects on us, yet in forming our judgments of them we are greatly at the mercy of circumstances, which either modify the impressions actually received, or combine them with adjuncts which have become habitually associated with different judgments; and, therefore, that, in estimating the degree of confidence we are to place in our conclusions, we must, of necessity, take into account these modifying or accompanying circumstances, whatever they may be. We do not, of course, here speak of deranged organization; such as, for instance, a distortion of the eye, producing double vision, and still less of mental delusion, which absolutely perverts the meaning of sensible impressions.

(74.) As the mind exists not in the place of sensible objects, and is not brought into immediate relation with them, we can only regard sensible impressions as signals conveyed from them by a wonderful, and, to us, inexplicable mechanism, to our minds, which receives and reviews them, and, by habit and association, connects them with corresponding qualities or affections in the objects; just as a person writing down and comparing the signals of a telegraph might interpret their meaning. As, for instance, if he had constantly observed that the exhibition of a certain signal was sure to be followed next day by the announcement of the arrival of a ship at Portsmouth, he would connect the two facts by a link of the very same nature with that which connects the notion of a large wooden building, filled with sailors, with the impression of her outline on the retina of a spectator on the beach.

(75.) In captain Head's amusing and vivid description of his journey across the Pampas of South America occurs an anecdote quite in point. His guide one day suddenly stopped him, and, pointing high into the air, cried out, "A lion!" Surprised at such an exclamation, accompanied with such an act, he turned up his eyes, and with difficulty perceived, at an immeasurable height, a flight of condors soaring in circles in a particular spot. Beneath that spot, far out of sight of himself or guide, lay the carcass of a horse, and over that carcass stood (as the guide well knew) the lion, whom the condors were eyeing with envy from their airy height The signal of the birds was to him what the sight of the lion alone could have been to the traveller, a full assurance of its existence.

CHAP. II.

OF THE ANALYSIS OF PHENOMENA.

(76.) *Phenomena*, then, or appearances, as the word is literally rendered, are the sensible results of processes and operations carried on among external objects, or their constituent principles, of which they are only signals, conveyed to our minds as aforesaid. Now, these processes themselves may be in many instances rendered *sensible;* that is to say, analysed, and shown to consist in the motions or other affections of sensible objects themselves. For instance, the phenomenon of the sound produced by a musical string, or a bell, when struck, may be shown to be the result of a process consisting in the rapid vibratory motion of its parts communicated to the air, and thence to our ears; though the immediate effect on our organs of hearing does not excite the least idea of such a motion. On the other hand, there are innumerable instances of sensible impressions which (at least at present) we are incapable of tracing beyond the mere sensation; for example, in the sensations of bitterness, sweetness, &c. These, accordingly, if we were inclined to form hasty decisions, might be regarded as ultimate qualities; but the instance of sounds, just adduced, alone would teach us caution in such decisions, and incline us to believe them mere results of some secret process going on in our organs of taste, which is too subtle for us to trace. A simple experiment will serve to set this in a clearer light. A solution of the salt called by chemists *nitrate of silver*, and another of the *hyposulphite of soda*, have each of them separately, when taken into the mouth, a disgustingly bitter taste; but if they be mixed, or if one be tasted before the mouth is thoroughly cleared of the other, the sensible impression is that of intense sweetness. Again, the salt called *tungstate of soda* when first tasted is sweet, but speedily changes to an intense and pure bitter, like quassia.*

(77.) How far we may ever be enabled to attain a knowledge of the ultimate and inward processes of nature in the production of phenomena, we have no means of knowing; but, to judge from the degree of obscurity which hangs about the only case in which we feel within ourselves a *direct* power to produce any one, there seems no great hope of penetrating so far. The case alluded to is the production of motion by the exertion of force. We are conscious of a power to move our limbs, and by their intervention other bodies; and that this effect is the result of a certain inexplicable process which we are aware of, but can no way describe in words, by which we exert *force*. And even when such exertion produces no visible effect, (as when we press our two hands violently together, so as just to oppose each other's effort,) we still perceive, by the fatigue and exhaustion, and by the impossibility of maintaining the effort long, that something is going on within us, of which the mind is the agent, and the will the determining cause. This impression

* Thomson's First Principles of Chemistry, vol. ii. p. 68.

which we receive of the nature of force, from our own effort and our sense of fatigue, is quite different from that which we obtain of it from seeing the effect of force exerted by others in producing *motion.* Were there no such thing as motion, had we been from infancy shut up in a dark dungeon, and every limb encrusted with plaster, this internal consciousness would give us a complete idea of *force;* but when set at liberty, habit alone would enable us to recognize its exertion by its *signal,* motion, and *that* only by finding that the same action of the mind which in our confined state enables us to fatigue and exhaust ourselves by the tension of our muscles, puts it in our power, when at liberty, to move ourselves and other bodies. But how obscure is our knowledge of the process going on within us in the exercise of this important privilege, in virtue of which alone we act as direct *causes,* we may judge from this, that when we put any limb in motion, the seat of the exertion seems to us to be *in* the limb, whereas it is demonstrably no such thing, but either in the brain or in the spinal marrow; the proof of which is, that if a little fibre, called a nerve, which forms a communication between the limb and the brain, or spine, be divided in any part of its course, however we may make the effort, the limb will not move.

(78.) This one instance of the obscurity which hangs about the only act of direct *causation* of which we have an immediate consciousness, will suffice to show how little prospect there is that, in our investigation of nature, we shall ever be able to arrive at a knowledge of ultimate causes, and will teach us to limit our views to that of *laws,* and to the analysis of complex phenomena by which they are resolved into simpler ones, which, appearing to us incapable of further analysis, we must consent to regard as causes. Nor let any one complain of this as a limitation of his faculties. We have here "ample room and verge enough" for the full exercise of all the powers we possess; and, besides, it does so happen, that we are actually able to trace up a very large portion of the phenomena of the universe to this one *cause,* viz. the exertion of mechanical *force;* indeed, so large a portion, that it has been made a matter of speculation whether this is not the only one that is capable of acting on material beings.

(79.) What we mean by the analysis of complex phenomena into simpler ones, will best be understood by an instance. Let us, therefore, take the phenomenon of sound, and, by considering the various cases in which sounds of all kinds are produced, we shall find that they all agree in these points:—1st, The excitement of a motion in the sounding body. 2dly, The communication of this motion to the air or other intermedium which is interposed between the sounding body and our ears. 3dly, The propagation of such motion from particle to particle of such intermedium in due succession. 4thly, Its communication, from the particles of the intermedium adjacent to the ear, to the ear itself. 5thly, Its conveyance in the ear, by a certain mechanism, to the auditory nerves. 6thly, The excitement of sensation. Now, in this analysis, we perceive that two principal matters must be understood, before we can have a true and complete knowledge of sound:—1st, The excitement and propagation of motion. 2dly, The production of sensation. These, then, are two other phenomena, of a simpler, or, it would be more correct to say, of a more general or elementary order, into which the complex phenomenon of sound resolves itself. But again, if we consider the communication of motion from body to body, or from one part to another of the same, we shall perceive that it is again resolvable into several other phenomena. 1st, The original setting in motion of a material body, or any part of one. 2dly, The behaviour of a particle set in motion, when it meets another lying in its way, or is otherwise impeded or influenced by its connection with surrounding particles. 3dly, The behaviour of the particles so impeding or influencing it under such circumstances; besides which, the last two point out another phenomenon, which it is necessary also to consider, viz. the phenomenon of the connection of the parts of material bodies in masses, by which they form aggregates, and are enabled to influence each other's motions.

(80.) Thus, then, we see that an analysis of the phenomenon of sound leads to the enquiry, 1st, of two *causes*, viz. the cause of motion, and the cause of sensation, these being phenomena which (at least as human knowledge stands at present) we are unable to analyse further; and, therefore, we set them down as simple, elementary, and referable, for any thing we can see to the contrary, to the immediate action of their causes. 2dly, Of several questions relating to the connection between the motion of material bodies and its cause, such as, *What will happen* when a moving body is surrounded on all sides by others not in motion? *What will happen* when a body not in motion is advanced upon by a moving one? It is evident that the answers to such questions as these can be no other than *laws of motion*, in the sense we have above attributed to laws of nature, viz. a statement in words of what will happen in such and such proposed general contingencies. Lastly, we are led, by pursuing the analysis, and considering the phenomenon of the aggregation of the parts of material bodies, and the way in which they influence each other, to two other general phenomena, viz., the cohesion and elasticity of matter; and these we have no means of analysing further, and must, therefore, regard them (till we see reasons to the contrary) as *ultimate phenomena*, and referable to the direct action of causes, viz. an attractive and a repulsive *force*.

(81.) Of force, as counterbalanced by opposing force, we have, as already said, an internal consciousness; and though it may seem strange to us that matter should be capable of exerting on matter the same kind of effort, which, judging alone from this consciousness, we might be led to regard as a mental one; yet we cannot refuse the direct evidence of our senses, which shows us that when we keep a spring stretched with one hand, we feel our effort opposed exactly in the same way as if we had ourselves opposed it with the other hand, or as it would be by that of another person. The enquiry, therefore, into the aggregation of matter resolves itself into the general question, What will be the behaviour of material particles under the mutual action of opposing forces capable of counterbalancing each other? and the answer to this question can be no other than the announcement of the *law* of equilibrium, whatever law that may be.

(82.) With regard to the cause of sensation, it must be regarded as much more obscure than that of motion, inasmuch as we have no conscious knowledge of it, *i. e.* we have no power, by any act of our minds and will, to call up a sensation. It is true, we are not destitute of an approach to it, since, by an effort of memory and imagination, we can produce in our minds an impression, or idea, of a sensation which, in peculiar cases, may even approach in vividness to actual reality. In dreams, too, and, in some cases of disordered nerves, we have sensations without objects. But if force, as a cause of motion, is obscure to us, even while we are in the act of exercising it, how much more so is this other cause, whose exercise we can only imitate imperfectly by any voluntary act, and of whose purely internal action we are only fully conscious when in a state that incapacitates us from reasoning, and almost from observation!

(83.) Dismissing, then, as beyond our reach, the enquiry into causes, we must be content at present to concentrate our attention on the laws which prevail among phenomena, and which seem to be their immediate results. From the instance we have just given, we may perceive that every enquiry into the intimate nature of a complex phenomenon branches out into as many different and distinct enquiries as there are simple or elementary phenomena into which it may be analysed; and that, therefore, it would greatly assist us in our study of nature, if we could, by any means, ascertain what *are* the ultimate phenomena into which all the composite ones presented by it may be resolved. There is, however, clearly no way by which this can be ascertained *a priori*. We must go to nature itself, and be guided by the same kind of rule as the chemist in his analysis, who accounts every ingredient an *element* till it can be decompounded and resolved into others. So, in natural philosophy, we must account every phenomenon an elementary or simple one till we can analyse it, and show that it is the result of others, which in their turn become elementary. Thus, in a modified and relative sense, we may still continue to speak of causes, not intending thereby those ultimate principles of action on whose exertion the whole frame of nature depends, but of those proximate links which connect phenomena with others of a simpler, higher, and more general or elementary kind. For example: we may regard the vibration of a musical string as the proximate cause of the sound it yields, receiving it, so far, as an ultimate fact, and waving or deferring enquiry into the cause of vibrations which is of a higher and more general nature.

(84.) Moreover, as in chemistry we are sometimes compelled to acknowledge the existence of elements different from those already identified and known, though we cannot insulate them, and to perceive that substances have the characters of compounds, and must therefore be susceptible of analysis, though we do not see how it is to be set about; so, in physics, we may perceive the complexity of a phenomenon, without being able to perform its analysis. For example: in magnetism, the agency of electricity is clearly made out, and they are shown to stand to one another in the relation of effect and cause, at least in so far as that all the phenomena of magnetism are producible by electricity, but no electric phenomena have hitherto ever been produced by magnetism. But the analysis of magnetism, in its relation to particular metals, is not yet quite satisfactorily performed; and we are compelled to admit the existence of some cause, whether proximate or ultimate, whose presence in the one and not in the other phenomenon determines their difference. Cases like these, of all which science presents, offer the highest interest. They excite enquiry, like the near approach to the solution of an enigma; they show us that there is light, could only a certain veil be drawn aside.

(85.) In pursuing the analysis of any phenomenon, the moment we find ourselves stopped by one of which we perceive no analysis, and which, therefore, we are forced to refer (at least provisionally) to the class of ultimate facts, and to regard as elementary, the study of that phenomenon and of its laws becomes a separate branch of science. If we encounter the same elementary phenomenon in the analysis of several composite ones, it becomes still more interesting, and assumes additional importance; while at the same time we acquire information respecting the phenomenon itself, by observing those with which it is habitually associated, that may help us at length to its analysis. It is thus that sciences increase, and acquire a mutual relation and dependency. It is thus, too, that we are at length enabled to trace parallels and analogies between great branches of science themselves, which at length terminate in a perception of their dependence on some common phenomenon of a more general and elementary nature than that which form the subject of either separately. It was thus, for example, that, previous to Oërsted's great discovery of electro-magnetism, a general resemblance between the two sciences of electricity and magnetism was recognised, and many of the chief phenomena in each were ascertained to have their parallels, *mutatis mutandis*, in the other. It was thus, too, that an analogy subsisting between sound and light has been gradually traced into a closeness of agreement, which can hardly leave any reasonable doubt of their ultimate coincidence in one common phenomenon, the vibratory motion of an elastic medium. If it be allowed to pursue our illustration from chemistry, and to ground its application not on what has been, but on what may one day be, done, it is thus that the general family resemblance between certain groups of bodies, now regarded as elementary, (as nickel and cobalt, for instance, chlorine, iode, and brome,) will, perhaps, lead us hereafter to perceive relations between them of a more intimate kind than we can at present trace.

(86.) On those phenomena which are most frequently encountered in an analysis of nature, and which most decidedly resist further decomposition, it is evident that the greatest pains and attention ought to be bestowed, not only because they furnish a key to the greatest number of enquiries, and serve to group and classify together the greatest range of phenomena, but by reason of their higher nature, and because it is in these that we must look for the direct action of causes, and the most extensive and general enunciation of the laws of nature. These, once discovered, place in our power the explanation of all particular facts, and become grounds of reasoning, independent of particular trial: thus playing the same part in natural philosophy that axioms do in geometry; containing, in a refined and condensed state, and as it were in a quintessence, all that our reason has occasion to draw from experience to enable it to follow out the truths of physics by the mere application of logical argument. Indeed, the axioms of geometry themselves may be regarded as in some sort an appeal to experience, not corporeal, but mental. When we say, the whole is greater than its part, we announce a general fact, which rests, it is true, on our ideas of whole and part; but, in abstracting these notions, we begin by considering them as subsisting in space, and time, and body, and again, in linear, and superficial, and solid space. Again, when we say, the equals of equals are equal, we mentally make comparisons, in equal spaces, equal times, &c.; so that these axioms, however self-evident, are still general propositions so far of the inductive kind, that, independently of experience, they would not present themselves to the mind.

The only difference between these and axioms ob tained from extensive induction is this, that, in raising the axioms of geometry, the instances offer themselves spontaneously, and without the trouble of search, and are few and simple; in raising those of nature, they are infinitely numerous, complicated, and remote; so that the most diligent research and the utmost acuteness are required to unravel their web, and place their meaning in evidence.

(87.) By far the most general phenomenon with which we are acquainted, and that which occurs most constantly, in every enquiry into which we enter, is motion, and its communication. Dynamics, then, or the science of force and motion, is thus placed at the head of all the sciences; and, happily for human knowledge, it is one in which the highest certainty is attainable, a certainty no way inferior to mathematical demonstration. As its axioms are few, simple, and in the highest degree distinct and definite, so they have at the same time an immediate relation to geometrical quantity, space, time, and direction, and thus accommodate themselves with remarkable facility to geometrical reasoning. Accordingly, their consequences may be pursued, by arguments purely mathematical. to any extent, insomuch that the limit of our knowledge of dynamics is determined only by that of pure mathematics, which is the case in no other branch of physical science.

(88.) But, it will now be asked, how we are to proceed to analyse a composite phenomenon into simpler ones, and whether any general rules can be given for this important process? We answer, None; any more than (to pursue the illustration we have already had recourse to) general rules can be laid down by the chemist for the analysis of substances of which all the ingredients are unknown. Such rules, could they be discovered, would include the whole of natural science; but we are very far, indeed, from being able to propound them. However, we are to recollect that the analysis of phenomena, philosophically speaking, is principally useful, as it enables us to recognize, and mark for special investigation, those which appear to us simple; to set methodically about determining their laws, and thus to facilitate the work of raising up general axioms, or forms of words, which shall include the whole of them; which shall, as it were, transplant them out of the external into the intellectual world, render them creatures of pure thought, and enable us to reason them out *à priori*. And what renders the power of doing this so eminently desirable is, that, in thus reasoning back from generals to particulars, the propositions at which we arrive apply to an immense multitude of combinations and cases, which were never individually contemplated in the mental process by which our axioms were first discovered; and that, consequently, when our reasonings are pushed to the utmost limit of particularity, their results appear in the form of *individual facts*, of which we might have had no knowledge from immediate experience; and thus we are not only furnished with the explanation of all known facts, but with the actual discovery of such as were before unknown. A remarkable example of this has already been mentioned in Fresnel's *à priori* discovery of the extraordinary refraction of both rays in a doubly refracting medium. To give another example: — The law of gravitation is a physical axiom of a very high and universal kind, and has been raised by a succession of inductions and abstractions drawn from the observation of numerous facts and subordinate laws in the planetary system. When this law is taken for granted, and laid down as a basis of reasoning, and applied to the actual condition of our own planet, one of the consequences to which it leads is, that the earth, instead of being an exact sphere, must be compressed or flattened in the direction of its polar diameter, the one diameter being about thirty miles shorter than the other; and this conclusion, deduced at first by mere reasoning, has been since found to be true in fact. All astronomical predictions are examples of the same thing.

(89.) In the important business of raising these axioms of nature, we are not, as in the analysis of phenomena, left wholly without a guide. The nature of abstract or general reasoning points out in a great measure the course we must pursue. A law of nature, being the statement of what will happen in certain general contingencies, may be regarded as the announcement, in the same words, of a whole group or class of phenomena. Whenever, therefore, we perceive that two or more phenomena agree in so many or so remarkable points, as to lead us to regard them as forming a class or group, if we lay out of consideration, or *abstract*, all the circumstances in which they disagree, and retain in our minds those only in which they agree, and then, under this kind of mental convention, frame a definition or statement of one of them, in such words that it shall apply equally to them all, such statement will appear in the form of a general proposition, having so far at least the character of a law of nature.

(90.) For example: a great number of transparent substances, when exposed, in a certain particular manner, to a beam of light which has been prepared by undergoing certain reflexions or refractions, (and has thereby acquired peculiar properties, and is said to be "*polarized*,") exhibit very vivid and beautiful colours, disposed in streaks, bands, &c. of great regularity, which seem to arise within the substance, and which, from a certain regular succession observed in their appearance, are called "periodical colours." Among the substances which exhibit periodical colours occur a great variety of transparent solids, but no fluids and no opaque solids. Here, then, there seems to be sufficient community of nature to enable us to use a general term, and to state the proposition as a law, viz. *transparent solids* exhibit periodical colours by exposure to polarized light. However, this, though true of many, does not apply to *all* transparent solids, and therefore we cannot state it as a general truth or law of nature in this form; although the reverse proposition, that all solids which exhibit such colours in such circumstances are *transparent*, would be correct and general. It becomes necessary, then, to make a list of those to which it does apply; and thus a great number of substances of all kinds become grouped together, in a class linked by this common property. If we examine the individuals of this group, we find among them the utmost variety of colour, texture, weight, hardness, form and composition; so that, in these respects, we seem to have fallen upon an assemblage of contraries. But when we come to examine them closely, in all their properties, we find they have all one point of agreement, in the property of double refraction, (see page 30.) and therefore we may describe them all truly as *doubly refracting substances*. We may, therefore, state the fact in the form, "Doubly refracting substances exhibit periodical colours by exposure to polarized light; and in this form it is found, on further examination to be true, not only for those particular instances which we had in view when we first propounded it, but in all cases which have since occurred on further enquiry, without a single exception; so that the proposition is general, and entitled to be regarded as a law of nature.

(91.) We may therefore regard a law of nature either, 1st, as a general proposition, announcing, in abstract terms, a whole group of particular facts relating to the behaviour of natural agents in proposed circumstances; or, 2dly, as a proposition announcing that a whole class of individuals agreeing in one character agree also in another. For example: in the case before us, the law arrived at includes, in its general announcement, among others, the particular facts, that rock crystal and saltpetre exhibit periodical colours; for these are both of them doubly refracting substances. Or, it may be regarded as announcing a relation between the two phenomena of double refraction, and the exhibition of periodical colours; which in the actual case is one of the most important, viz. the relation of *constant association,* inasmuch as it asserts that in whatever individual the one character is found, the other will invariably be found also.

(92.) These two lights, in which the announcement of a general law may be regarded, though at bottom they come to the same thing, yet differ widely in their influence on our minds. The former exhibits a law as little more than a kind of artificial memory; but in the latter it becomes a step in philosophical investigation, leading directly to the consideration of a proximate, if not an ultimate, cause; inasmuch as, whenever two phenomena are observed to be invariably connected together, we conclude them to be related to each other, either as cause and effect, or as common effects of a single cause.

(93.) There is still another light in which we may regard a law of the kind in question, viz. as a proposition asserting the mutual connection, or in some cases the entire identity, of two classes of individuals (whether individual objects or individual facts); and this is, perhaps, the simplest and most instructive way in which it can be conceived, and that which furnishes the readiest handle to further generalization in the raising of yet higher axioms. For example: in the case above mentioned, if observation had enabled us to establish he existence of a class of bodies possessing the property of double refraction, and observations of another kind had, independently of the former, led us to recognize a class possessing that of the exhibition of periodical colours in polarized light, a mere comparison of lists would at once demonstrate the identity of the two classes, or enable us to ascertain whether one was or was not included in the other.

(94.) It is thus we perceive the high importance in physical science of just and accurate classifications of particular facts, or individual objects, under general well considered heads or points of agreement (for which there are none better adapted than the simple phenomena themselves into which they can be analysed in the first instance); for by so doing each of such phenomena, or heads of classification, becomes not a particular but a general fact; and when we have amassed a great store of such *general facts*, they become the objects of another and higher species of classification, and are themselves included in laws which, as they dispose of groups, not individuals, have a far superior degree of generality, till at length, by continuing the process, we arrive at *axioms* of the highest degree of generality of which science is capable.

(95.) This process is what we mean by induction; and, from what has been said, it appears that induction may be carried on in two different ways,—either by the simple juxta-position and comparison of ascertained classes, and marking their agreements and disagreements; or by considering the individuals of a class, and casting about, as it were to find in what particular they all agree, besides that which serves as their principle of classification. Either of these methods may be put in practice as one or the other may afford facilities in any case; but it will naturally happen that, where facts are numerous, well observed, and methodically arranged, the former will be more applicable than in the contrary case: the one is better adapted to the maturity, the other to the infancy, of science: the one employs, as an engine, the division of labour; the other mainly relies on individual penetration, and requires a union of many branches of knowledge in one person.

CHAP. III.

OF THE STATE OF PHYSICAL SCIENCE IN GENERAL, PREVIOUS TO THE AGE OF GALILEO AND BACON.

(96.) It is to our immortal countryman Bacon that we owe the broad announcement of this grand and fertile principle; and the developement of the idea, that the whole of natural philosophy consists entirely of a series of inductive generalizations, commencing with the most circumstantially stated particulars, and carried up to universal laws, or axioms, which comprehend in their statements every subordinate degree of generality, and of a corresponding series of inverted reasoning from generals to particulars, by which these axioms are traced back into their remotest consequences, and all particular propositions deduced from them; as well those by whose immediate consideration we rose to their discovery, as those of which we had no previous knowledge. In the course of this descent to particulars, we must of necessity encounter all those facts on which the arts and works that tend to the accommodation of human life depend, and acquire thereby the command of an unlimited practice, and a disposal of the powers of nature co-extensive with those powers themselves. A noble promise, indeed, and one which ought, surely, to animate us to the highest exertion of our faculties; especially since we have already such convincing proof that it is neither vain nor rash, but, on the contrary, has been, and continues to be, fulfilled, with a promptness and liberality which even its illustrious author in his most sanguine mood would have hardly ventured to anticipate.

(97.) Previous to the publication of the Novum Organum of Bacon, natural philosophy, in any legitimate and extensive sense of the word, could hardly be said to exist. Among the Greek philosophers, of whose attainments in science alone, in the earlier ages of the world, we have any positive knowledge, and that but a very limited one, we are struck with the remarkable contrast between their powers of acute and subtle disputation, their extraordinary success in abstract reasoning, and their intimate familiarity with subjects purely intellectual, on the one hand; and, on the other, with their loose and careless consideration of external nature, their grossly illogical deductions of principles of sweeping generality from few and ill-observed facts, in some cases; and their reckless assumption of abstract principles having no foundation but in their own imaginations, in others; mere forms of words, with nothing corresponding to them in nature, from which, as from mathematical definitions, postulates, and axioms, they imagined that all phenomena could be derived, all the laws of nature deduced. Thus, for instance, having settled it in their own minds, that a circle is the most perfect of figures, they concluded, of course, that the movements of the heavenly bodies must all be performed in exact circles, and with uniform motions; and when the plainest observation demonstrated the contrary, instead of doubting the principle, they saw no better way of getting out of the difficulty than by having recourse to endless combinations of circular motions to preserve their ideal perfection.

(98.) Undoubtedly among the Greek philosophers were many men of transcendent talents and virtues, the ornaments of their species, and justly entitled to the veneration of all posterity; but regarded as a body they can hardly be considered otherwise than as a knot of disputatious candidates for popular favour, too busy in maintaining their ascendency over their followers and admirers, by an ostentatious display of superior knowledge, to have the leisure (had they always the inclination) to base their pretensions on a deep and sure foundation, and yet too sensible of the disgrace and inconvenience of failure, not to defend their dogmas, however shallow, when once promulgated, against their keen and sagacious opponents, by every art of sophism or appeal to passion. Hence the crudities and chimerical views with which their systems of philosophy, both natural and moral, were overloaded; their endless disputes about verbal subtleties, and, last and worst, the proud assumption with which they sheltered ignorance and indolence under the screen of unintelligible jargon or dogmatical assertion. Perhaps, however, this character applies rather to the later than to the earlier of the Greek philosophers. The spirit of rational enquiry into nature seems, if we can judge from the uncertain and often contradictory notices handed down to us of their tenets, to have been far more alive, and less warped by this vain and arrogant turn, then than at a later period. We know not now what was the precise meaning attached by Thales to his opinion, that water was the origin of all things; but modern geologists will not be at a loss to conceive how an observant traveller might become impressed with this notion, without having recourse to the mystic records of Egypt or Chaldea. His ideas of eclipses and of the nature of the moon were sound; and his prediction of an eclipse of the sun, in particular, was attended with circumstances so remarkable as to have made it a matter of important investigation to modern astronomers. Anaxagoras, among a number of crude and imperfectly explained notions, speculated rationally enough on the cause of the winds and of the rainbow, and less absurdly on earthquakes than many modern geologists have done, and appears generally to have had his attention alive to nature, and his mind open to just reasoning on its phenomena; while Pythagoras, whether he reasoned it out for himself, or borrowed the notion from Egypt or India, had attained a just conception of the general disposition of the parts of the solar system, and the place held by the earth in it; nay, according to some accounts, had even raised his views so far as to speculate on the attraction of the sun as the bond of its union.

(99.) But the successors of these *bonâ fide* enquirers into nature debased the standard of truth; and, taking advantage of the credit justly attached to their discoveries, renounced the modest character of learners, and erected themselves into teachers, and, to maintain their pretensions to this character, adopted the tone of men who had nothing further to learn. Unfortunately for true science, the national character gave every encouragement to pretensions of this kind. That restless craving after novelty, which distinguished the Greeks in their civil and political relations, pursued them into their philosophy. Whatever speculations were only ingenious and new had irresistible charms; and the teacher who could embody a clever thought in elegant language, or at once save his followers and himself the trouble of thinking or reasoning, by bold assertion, was too often induced to acquire cheaply the reputation of superior knowledge, snatch a few superficial notions from the most ordinary and obvious facts, envelope them in a parade of abstruse words, declare them the primary and ultimate principles of all things, and denounce as absurd and impious all opinions opposed to his own.

(100.) In this war of words the study of nature was neglected, and an humble and patient enquiry after facts altogether despised, as unworthy of the high *priori* ground a true philosopher ought to take. It was the radical error of the Greek philosophy to imagine that the same method which proved so eminently successful in mathematical, would be equally so in physical, enquiries, and that, by setting out from a few simple and almost self-evident notions, or *axioms*, every thing could be reasoned out. Accordingly, we find them constantly straining their invention to discover these principles, which were to prove so pregnant. One makes *fire* the essential matter and origin of the universe; another, *air*; a third, discovers the key to every difficulty, and the explanation of all phenomena, in the "το απειρον" or infinitude of things; a fourth, in the το ὀν and the το μη ὀν, that is to say, in entity and nonentity;—

till at length an authority, which was destined to command opinions for nearly two thousand years, settled this important point, by deciding, that *matter form*, and *privation*, were to be considered the principles of all things.

(101.) It were to do injustice to Aristotle, however, to judge of him by *such* a sample of his philosophy He, at least, saw the necessity of having recourse to nature for something like principles of physical science; and, as an observer, a collector and recorder of facts and phenomena, stood without an equal in his age. It was the fault of that age, and of the perverse and flimsy style of verbal disputation which had infected all learning, rather than his own, that he allowed himself to be contented with vague and loose notions drawn from general and vulgar observation, in place of seeking carefully, in well arranged and thoroughly considered instances, for the true laws of nature. His voluminous works, on every department of human knowledge existing in his time, have nearly all perished. From his work on animals, which has descended to us, we are, however, enabled to appreciate his powers of observation; and a parallel drawn by an eminent Oxford professor between his classifications and those of the most illustrious of living naturalists, shows him to have attained a view of animated nature in a remarkable degree comprehensive, and which contrasts strikingly with the confusion, vagueness, and assumption of his physical opinions and dogmas. In these it is easy to recognize a mind not at home, and an impression of the necessity of saying something learned and systematic, without knowing what to say. Thus he divides motions into natural and unnatural; the natural motion of fire and light bodies being upwards, those of heavy downwards, each seeking its kindred nature in the heavens and the earth. Thus, too, the immediate impressions made on us by external objects, such as hardness, colour, heat, &c. are referred at once, in the Aristotelian philosophy, to occult qualities, in virtue of which they are as they are, and beyond which it is useless to enquire.*

* Galileo exposes unsparingly the Aristotelian style of reasoning. The reader may take the following from him as a specimen of its quality. The object is to prove the immutability and incorruptibility of the heavens; and thus it is done:—

I. Mutation is either generation or corruption.

II. Generation and corruption only happen between contraries.

III. The motions of contraries are contrary.

IV. The celestial motions are circular.

V. Circular motions have no contraries.

α. Because there can be but three simple motions.

1. To a centre.
2. Round a centre.
3. From a centre.

β. Of three things, one only can be contrary to one.

γ. But a motion to a centre is manifestly the contrary to a motion from a centre.

δ. Therefore a motion *round* a centre (*i. e.* a circular motion) remains without a contrary.

VI. *Therefore* celestial motions have no contraries — *therefore* among celestial *things* there are no contraries — *therefore* the heavens are eternal, immutable, incorruptible, and so forth.

It is evident that all this string of nonsense depends on the excessive vagueness of the notions of generation, corruption, contrariety, &c. on which the changes are rung. — *See* GALILEO, *Systema Cosmicum*, Dial. i. p. 30.

Of course there will occur a limit beyond which it *is* useless for merely human faculties to enquire; but where that limit is placed, experience alone can teach us; and at least to assert that we *have* attained it, is now universally recognized as the sure criterion of dogmatism.

(102.) In the early ages of the church the writings of Aristotle were condemned, as allowing too much to reason and sense; and even so late as the twelfth century they were sought out and burned, and their readers excommunicated. By degrees, however, the extreme injustice of this impeachment of their character was acknowledged: they became the favourite study of the schoolmen, and furnished the keenest weapons of their controversy, being appealed to in all disputes as of sovereign authority; so that the slightest dissent from any opinion of the "great master," however absurd or unintelligible, was at once drowned by clamour, or silenced by the still more effectual argument of bitter persecution. If the logic of that gloomy period could be justly described as "the art of talking unintelligibly on matters of which we are ignorant," its physics might, with equal truth, be summed up in a deliberate preference of ignorance to knowledge, in matters of every day's experience and use.

(103.) In "this opake of nature and of soul," the perverse activity of the alchemists from time to time struck out a doubtful spark*; and our

* Macquer justly observes, that the alchemists would have rendered essential service to chemistry had they only related their unsuccessful experiments as clearly as they have obscurely related those which they pretend to have been successful.—*Macquer's Dictionary of Chemistry*, i. x.

illustrious countryman, Roger Bacon, shone out at the obscurest moment, like an early star predicting dawn. It was not, however, till the sixteenth century that the light of nature began to break forth with a regular and progressive increase. The vaunts of Paracelsus of the power of his chemical remedies and elixirs, and his open condemnation of the ancient pharmacy, backed as they were by many surprising cures, convinced all rational physicians that chemistry could furnish many excellent remedies, unknown till that time*, and a number of valuable experiments began to be made by physicians and chemists, desirous of discovering and describing new chemical remedies. The chemical and metallurgic arts exercised by persons empirically acquainted with their secrets, began to be seriously studied with a view to the acquisition of rational and useful knowledge, and regular treatises on branches of natural science at length to appear. George Agricola, in particular, devoted himself with ardour to the study of mineralogy and metallurgy in the mining districts of Bohemia and Schemnitz, and published copious and methodical accounts of all the facts within his knowledge: and our countryman, Dr. Gilbert of Colchester, in

* Paracelsus performed most of these cures by mercury and opium, the use of which latter drug he had learned in Turkey. Of mercurial preparations the physicians of his time were ignorant, and of opium they were afraid, as being "cold in the fourth degree." Tartar was likewise a great favourite of Paracelsus, who imposed on it that name, "because it contains the water, the salt, the oil, and the acid, which burn the patient as hell does:" in short, a kind of counterbalance to his opium.

1590, published a treatise on magnetism, full of valuable facts and experiments, ingeniously reasoned on; and he likewise extended his enquiries to a variety of other subjects, in particular to electricity.

(104.) But, as the decisive mark of a great commencing change in the direction of the human faculties, astronomy, the only science in which the ancients had made any real progress, and ascended to any thing like large and general conceptions, began once more to be studied in the best spirit of a candid philosophy; and the Copernican or Pythagorean system arose or revived, and rapidly gained advocates. Galileo at length appeared, and openly attacked and refuted the Aristotelian dogmas respecting motion, by direct appeal to the evidence of sense, and by experiments of the most convincing kind. The persecutions which such a step drew upon him, the record of his perseverance and sufferings, and the ultimate triumph of his opinions and reasonings, have been too lately and too well related* to require repetition here.

(105.) By the discoveries of Copernicus, Kepler, and Galileo, the errors of the Aristotelian philosophy were effectually overturned on a plain appeal to the facts of nature; but it remained to show on broad and general principles, how and why Aristotle was in the wrong; to set in evidence the peculiar weakness of his method of philosophizing, and to substitute in its place a stronger and better. This

* See the Life of Galileo Galilei, by Mr. Drinkwater, with Illustrations of the Advancement of Experimental Philosophy.

important task was executed by Francis Bacon, Lord Verulam, who will, therefore, justly be looked upon in all future ages as the great reformer of philosophy, though his own actual contributions to the stock of physical truths were small, and his ideas of particular points strongly tinctured with mistakes and errors, which were the fault rather of the general want of physical information of the age than of any narrowness of view on his own part; and of this he was fully aware. It has been attempted by some to lessen the merit of this great achievement, by showing that the inductive method had been practised in many instances, both ancient and modern, by the mere instinct of mankind; but it is not the introduction of inductive reasoning, as a new and hitherto untried process, which characterises the Baconian philosophy, but his keen perception, and his broad and spirit-stirring, almost enthusiastic, announcement of its paramount importance, as the alpha and omega of science, as the grand and only chain for the linking together of physical truths, and the eventual key to every discovery and every application. Those who would deny him his just glory on such grounds would refuse to Jenner or to Howard their civic crowns, because a few farmers in a remote province had, time out of mind, been acquainted with vaccination, or philanthropists, in all ages, had occasionally visited the prisoner in his dungeon.

(106.) An immense impulse was now given to science, and it seemed as if the genius of mankind, long pent up, had at length rushed eagerly upon Nature, and commenced, with one accord, the great work of turning up her hitherto unbroken soil, and exposing the treasures so long concealed. A general sense now prevailed of the poverty and insufficiency of existing knowledge in *matters of fact;* and, as information flowed fast in, an era of excitement and wonder commenced, to which the annals of mankind had furnished nothing similar. It seemed, too, as if Nature herself seconded the impulse; and, while she supplied new and extraordinary aids to those senses which were henceforth to be exercised in her investigation, — while the telescope and the microscope laid open *the infinite* in both directions, — as if to call attention to her wonders, and signalise the epoch, she displayed the rarest, the most splendid and mysterious, of all astronomical phenomena, the appearance and subsequent total extinction of a new and brilliant fixed star twice within the lifetime of Galileo himself. *

(107.) The immediate followers of Bacon and Galileo ransacked all nature for new and surprising facts, with something of that craving for the marvellous, which might be regarded as a remnant of the age of alchemy and natural magic, but which, under proper regulation, is a most powerful and useful stimulus to experimental enquiry. Boyle, in particular, seemed animated by an enthusiasm of ardour, which hurried him from subject to subject,

* The temporary star in Cassiopeia observed by Cornelius Gemma, in 1572, was so bright as to be seen at noon-day. That in Serpentarius, first seen by Kepler in 1604, exceeded in brilliancy all the other stars and planets.

and from experiment to experiment, without a moment's intermission, and with a sort of undistinguishing appetite; while Hooke (the great contemporary, and almost the worthy rival, of Newton) carried a keener eye of scrutinising reason into a range of research even yet more extensive. As facts multiplied, leading phenomena became prominent, laws began to emerge, and generalizations to commence; and so rapid was the career of discovery, so signal the triumph of the inductive philosophy, that a single generation and the efforts of a single mind sufficed for the establishment of the system of the universe, on a basis never after to be shaken.

(108.) We shall now endeavour to enumerate and explain in detail the principal steps by which legitimate and extensive inductions are arrived at, and the processes by which the mind, in the investigation of natural laws, purges itself by successive degrees of the superfluities and incumbrances which hang about particulars, and obscure the perception of their points of resemblance and connection. We shall state the helps which may be afforded us, in a work of so much thought and labour, by a methodical course of proceeding, and by a careful notice of those means which have at any time been found successful, with a view to their better understanding and adaptation to other cases: a species of mental induction of no mean utility and extent in itself; inasmuch as by pursuing it alone we can attain a more intimate knowledge than we actually possess of the laws which regulate our discovery of truth, and of the rules, so far as they extend, to which invention is reducible. In doing this, we shall commence at the beginning, with experience itself, considered as the accumulation of the knowledge of individual objects and facts.

CHAP. IV.

OF THE OBSERVATION OF FACTS AND THE COLLECTION OF INSTANCES.

(109.) NATURE offers us two sorts of subjects of contemplation in the external world, — objects, and their mutual actions. But, after what has been said on the subject of sensation, the reader will be at no loss to perceive that we know nothing of the objects themselves which compose the universe, except through the medium of the impressions they excite in us, which impressions are the results of certain actions and processes in which sensible objects and the material parts of ourselves are directly concerned. Thus, our observation of external nature is limited to the mutual action of material objects on one another; and to facts, that is, the associations of phenomena or appearances. We gain no information by perceiving merely that an object is black; but if we also perceive it to be fluid, we at least acquire the knowledge that blackness is not incompatible with fluidity, and have thus made a step, however trifling, to a knowledge of the more intimate nature of these two qualities. Whenever, therefore, we would either analyse a phenomenon into simpler ones, or ascertain what is the course or law of nature under any proposed general contingency, the first step is to accumulate a sufficient quantity of well ascertained facts, or recorded instances, bearing on the point in question. Common sense dictates this, as affording us the means of examining the same subject in several points of view; and it would also dictate, that the more different these collected facts are in all other circumstances but that which forms the subject of enquiry, the better; because they are then in some sort brought into contrast with one another in their points of disagreement, and thus tend to render those in which they agree more prominent and striking.

(110.) The only facts which can ever become useful as grounds of physical enquiry are those which happen uniformly and invariably under the same circumstances. This is evident: for if they have not this character they cannot be included in laws; they want that universality which fits them to enter as elementary particles into the constitution of those universal axioms which we aim at discovering. If one and the same result does not constantly happen under a given combination of circumstances, apparently the same, one of two things must be supposed,—caprice (*i.e.* the arbitrary intervention of mental agency), or differences in the circumstances themselves, really existing, but unobserved by us. In either case, though we may record such facts as curiosities, or as awaiting explanation when the difference of circumstances shall be understood, we can make no use of them in scientific enquiry. Hence, whenever we notice a remarkable effect of any kind, our first question ought to be, Can it be reproduced? What are the circumstances under which it has happened? And will it *always* happen again if those circumstances, so far as we have been able to collect them, co-exist?

(111.) The circumstances, then, which accompany any observed fact, are main features in its observation, at least until it is ascertained by sufficient experience what circumstances have nothing to do with it, and might therefore have been left unobserved without sacrificing *the fact.* In observing and recording a fact, therefore, altogether new, we ought not to omit any circumstance capable of being noted, lest some one of the omitted circumstances should be essentially connected with the fact, and its omission should, therefore, reduce the implied statement of a *law of nature* to the mere record of an *historical event.* For instance, in the fall of meteoric stones, flashes of fire are seen proceeding from a cloud, and a loud rattling noise like thunder is heard. These circumstances, and the sudden stroke and destruction ensuing, long caused them to be confounded with an effect of lightning, and called thunderbolts. But one circumstance is enough to mark the difference: the flash and sound have been perceived occasionally to emanate from a *very small cloud* insulated in *a clear sky;* a combination of circumstances which never happens in a thunder storm, but which is undoubtedly intimately connected with their real origin.

(112.) Recorded observation consists of two distinct parts: 1st, an exact notice of the thing observed, and of all the particulars which may be supposed to have any natural connection with it; and, 2dly, a true and faithful record of them. As our senses are the only inlets by which we receive impressions of facts, we must take care, in observing, to have them all in activity, and to let nothing escape notice which affects any one of them. Thus, if lightning were to strike the house we inhabit, we ought to notice what kind of light we saw—whether a sheet of flame, a darting spark, or a broken zig-zag; in what direction moving, to what objects adhering, its colour, its duration, &c.; what sounds were heard—explosive, crashing, rattling, momentary, or gradually increasing and fading, &c.; whether any smell of fire was perceptible, and if sulphureous, metallic, or such as would arise merely from substances scorched by the flash, &c.; whether we felt any shock, stroke, or peculiar sensation, or experienced any strange taste in our mouths. Then, besides detailing the effects of the stroke, all the circumstances which might in any degree seem likely to attract, produce, or modify it, such as the presence of conductors, neighbouring objects, the state of the atmosphere, the barometer, thermometer, &c., and the disposition of the clouds, should be noted; and after all this particularity, the question *how* the house *came to be struck?* might ultimately depend on the fact that a flash of lightning twenty miles off passed at that particular moment *from the ground to the clouds*, by an effect of what has been termed the returning stroke.

(113.) A writer in the Edinburgh Philosophical Journal* states himself to have been led into a series of investigations on the chemical nature of a peculiar acid, by noticing, accidentally, a bitter taste in a

* Edinburgh Phil. Journ. 1819, vol. i. p. 8.

liquid about to be thrown away. Chemistry is full of such incidents.

(114.) In transient phenomena, if the number of particulars be great, and the time to observe them short, we must consult our memory before they have had time to fade, or refresh it by placing ourselves as nearly as possible in the same circumstances again; go back to the spot, for instance, and try the words of our statement by appeal to all remaining indications, &c. This is most especially necessary where we have not observed ourselves, but only collect and record the observations of others, particularly of illiterate or prejudiced persons, on any rare phenomenon, such as the passing of a great meteor, — the fall of a stone from the sky, — the shock of an earthquake, — an extraordinary hail-storm, &c.

(115.) In all cases which admit of numeration or measurement, it is of the utmost consequence to obtain precise numerical statements, whether in the measure of time, space, or quantity of any kind. To omit this, is, in the first place, to expose ourselves to illusions of sense which may lead to the grossest errors. Thus, in alpine countries, we are constantly deceived in heights and distances; and when we have overcome the first impression which leads us to under-estimate them, we are then hardly less apt to run into the opposite extreme. But it is not merely in preserving us from exaggerated impressions that numerical precision is desirable. It is the very soul of science; and its attainment affords the only criterion, or at least the best, of the truth of theories, and the correctness of experiments. Thus, it was entirely to the omission of exact numerical determinations of quantity that the mistakes and confusion of the Stahlian chemistry were attributable,—a confusion which dissipated like a morning mist as soon as precision, in this respect, came to be regarded as essential. Chemistry is in the most pre-eminent degree a science of quantity; and to enumerate the discoveries which have arisen in it, from the mere determination of weights and measures, would be nearly to give a synopsis of this branch of knowledge. We need only mention the law of definite proportions, which fixes the composition of every body in nature in determinate proportional weights of its ingredients.

(116.) Indeed, it is a character of all the higher laws of nature to assume the form of precise *quantitative* statement. Thus, the law of gravitation, the most universal truth at which human reason has yet arrived, expresses not merely the general fact of the mutual attraction of all matter; not merely the vague statement that its influence decreases as the distance increases, but the exact numerical rate at which that decrease takes place; so that when its amount is known at any one distance it may be calculated exactly for any other. Thus, too, the laws of crystallography, which limit the forms assumed by natural substances, when left to their own inherent powers of aggregation, to precise geometrical figures, with fixed angles and proportions, have the same essential character of strict mathematical expression, without which no exact particular conclusions could ever be drawn from them.

(117.) But, to arrive at laws of this description, it is

evident that every step of our enquiry must be perfectly free from the slightest degree of looseness and indecision, and carry with it the full force of strict numerical announcement; and that, therefore, the observations themselves on which all laws ultimately rest ought to have the same property. None of our senses, however, gives us direct information for the exact comparison of quantity. Number, indeed, that is to say, integer number, is an object of sense, because we can count; but we can neither weigh, measure, nor form any precise estimate of fractional parts by the unassisted senses. Scarcely any man could tell the difference between twenty pounds and the same weight increased or diminished by a few ounces; still less could he judge of the proportion between an ounce of gold and a hundred grains of cotton by balancing them in his hands. To take another instance: the eye is no judge of the proportion of different degrees of illumination, even when seen side by side; and if an interval elapses, and circumstances change, nothing can be more vague than its judgments. When we gaze with admiration at the gorgeous spectacle of the golden clouds at sunset, which seem drenched in light and glowing like flames of real fire, it is hardly by any effort we can persuade ourselves to regard them as the very same objects which at noonday pass unnoticed as mere white clouds basking in the sun, only participating, from their great horizontal distance, in the ruddy tint which luminaries acquire by shining through a great extent of the vapours of the atmosphere, and thereby even losing something of their light. So it is with our estimates of time, velocity, and all other matters of quantity; they are absolutely vague, and inadequate to form a foundation for any exact conclusion.

(118.) In this emergency we are obliged to have recourse to instrumental aids, that is, to contrivances which shall substitute for the vague impressions of sense the precise one of number, and reduce all measurement to counting. As a first preliminary towards effecting this, we fix on convenient *standards* of weight, dimension, time, &c., and invent contrivances for readily and correctly repeating them as often as we please, and counting how often such a standard unit is contained in the thing, be it weight, space, time, or angle, we wish to measure; and if there be a fractional part over, we measure this as a new quantity by aliquot parts of the former standard.

(119.) If every scientific enquirer observed only for his own satisfaction, and reasoned only on his own observations, it would be of little importance what standards he used, or what contrivances (if only just ones) he employed for this purpose; but if it be intended (as it is most important they should) that observations once made should remain as records to all mankind, and to all posterity, it is evidently of the highest consequence that all enquirers should agree on the use of a common standard, and that this should be one not liable to change by lapse of time. The selection and verification of such standards, however, will easily be understood to be a matter of extreme difficulty, if only from the mere circumstance that, to verify the permanence of one standard, we must compare it with others, which it is possible may be themselves inaccurate, or, at least, stand in need of verification.

(120.) Here we can only call to our assistance the presumed permanence of the great laws of nature, with all experience in its favour, and the strong impression we have of the general composure and steadiness of every thing relating to the gigantic mass we inhabit—"the great globe itself." In its uniform rotation on its axis, accordingly, we find a standard of time, which nothing has ever given us reason to regard as subject to change, and which, compared with other periods which the revolutions of the planets about the sun afford, has demonstrably undergone none since the earliest history. In the dimensions of the earth we find a natural unit of the measure of space, which possesses in perfection every quality that can be desired; and in its attraction combined with its rotation the researches of dynamical science have enabled us, through the medium of the pendulum, to obtain another invariable standard, more refined and less obvious, it is true, in its origin, but possessing a great advantage in its capability of ready verification, and therefore easily made to serve as a check on the other. The former, viz. direct measurement of the dimensions of the earth, is the origin of the *mètre*, the French unit of linear measure; the latter, of the British yard. Theoretically speaking, they are equally eligible; but when we consider that the *quantity directly measured*, in the case of the mètre, is a length a great many thousand times the final unit, and in the pendulum or yard very nearly the unit itself, there can be no hesitation in giving the preference as an original measure to the former, because any error committed in the process by which that is determined becomes subdivided in the final result; while, on the other hand, any minute error committed in determining the length of the pendulum becomes multiplied by the repetition of the unit in all measurements of considerable lengths performed in yards.

(121.) The same admirable invention of the pendulum affords a means of subdividing time to an almost unlimited nicety. A clock is nothing more than a piece of mechanism for counting the oscillations of a pendulum; and by that peculiar property of the pendulum, that one vibration commences exactly where the last terminates, no part of time is lost or gained in the juxta-position of the units so counted, so that the precise fractional part of a day can be ascertained which each such unit measures.

(122.) It is owing to this peculiar property by which the *juxta-position* of units of time and weight can be performed *without error*, that the whole of the accuracy with which time and weight can be multiplied and subdivided is owing. * The same thing cannot be accomplished in *space*, by any method

* The abstract principle of repetition in matters of measurement (viz. juxta-position of units without error) is applicable to a great variety of cases in which quantities are required to be determined to minute nicety. In chemistry, in determining the standard atomic weights of bodies, it seems easily and completely applicable, by a process which will suggest itself at once to every chemist, and seems the only thing wanting to place the exactness of chemical determinations on a par with astronomical measurements.

we are yet acquainted with, so that our means of subdividing space are much inferior in precision. The beautiful principle of repetition, invented by Borda, offers the nearest approach to it, but cannot be said to be absolutely free from the source of error in question. The method of "double weighing," which we owe to the same distinguished observer, affords an instance of the direct comparison of two equal weights independent of almost every source of error which can affect the comparison of one object with another. It has been remarked by Biot, that previous to the invention of this elegant method, instruments afforded no perfect means of ascertaining the weight of a body.

(123.) But it is not enough to possess a standard of this abstract kind: a real material measure must be constructed, and exact copies of it taken. This, however, is not very difficult; the great difficulty is to preserve it unaltered from age to age; for unless we transmit to posterity the units of our measurements, *such as we have ourselves used them*, we, in fact, only half bequeath to them our observations. This is a point too much lost sight of, and it were much to be wished that some direct provision for so important an object were made.*

* Accurate and *perfectly* authentic copies of the yard and pound, executed in platina, and hermetically sealed in glass, should be deposited deep in the interior of the massive stone-work of some great public building, whence they could only be rescued with a degree of difficulty sufficient to preclude their being disturbed unless on some very high and urgent occasion. The fact should be publicly recorded, and its memory preserved by an inscription. Indeed, how much valuable and useful information

(124.) But, it may be asked, if our measurement of quantity is thus unavoidably liable to error, how is it possible that our observations can possess that quality of numerical veracity which is requisite to render them the foundation of laws, whose distinguishing perfection consists in their strict mathematical expression? To this the reply is twofold. 1st, that though we admit the necessary existence of numerical error in every observation, we can always assign a limit which such error cannot possibly exceed; and the extent of this *latitude of error of observation* is less in proportion to the perfection of the instrumental means we possess, and the care bestowed on their employment. In the greater part of modern measurements it is, in point of fact, extremely minute, and may be still further diminished, almost to any required extent, by repeating the measurements a great number of times, and under a great variety of circumstances, and taking a mean of the results, when errors of opposite kinds will, at length, compensate each other. But, 2dly, there

---

of the actual existing state of arts and knowledge at any period might be transmitted to posterity in a distinct, tangible, and imperishable form, if, instead of the absurd and useless deposition of a few coins and medals under the foundations of buildings, specimens of ingenious implements or condensed statements of scientific truths, or processes in arts and manufactures, were substituted. Will books infallibly preserve to a remote posterity all that we may desire should be hereafter known of ourselves and our discoveries, or all that posterity would wish to know? and may not a useless ceremony be thus transformed into an act of enrolment in a perpetual archive of what we most prize, and acknowledge to be most valuable?

exists a much more fundamental reply to this objection. In reasoning upon our observations, the existence and possible amount of quantitative error is always to be allowed for; and the extent to which theories may be affected by it is never to be lost sight of. In reasoning upwards, from observations confessedly imperfect to general laws, we must take care always to regard our conclusions as conditional, so far as they may be affected by such unavoidable imperfections; and when at length we shall have arrived at our highest point, and attained to axioms which admit of general and deductive reasoning, the question, whether they *are* vitiated by the errors of observation or not, will still remain to be decided, and must become the object of subsequent verification. This point will be made the subject of more distinct consideration hereafter, when we come to speak of the verification of theories and the laws of probability.

(125.) With respect to our record of observations, it should be not only circumstantial but *faithful;* by which we mean, that it should contain all we did *observe*, and nothing else. Without any intention of falsifying our record, we may do so unperceived by ourselves, owing to a mixture of the views and language of an erroneous theory with that of simple fact. Thus, for example, if, in describing the effect of lightning, we should say, "The thunderbolt struck with violence against the side of the house, and beat in the wall," a fact would be stated which we did not see, and would lead our hearers to believe that a solid or ponderable projectile was concerned. The "strong smell of sulphur." which is sometimes said to accompany lightning, is a remnant of the theory which made thunder and lightning the explosion of a kind of aërial gunpowder, composed of sulphureous and nitrous exhalations. There are some subjects particularly infested with this mixture of theory in the statement of observed fact. The older chemistry was so overborne by this mischief, as quite to confound and nullify the descriptions of innumerable curious and laborious experiments. And in geology, till a very recent period, it was often extremely difficult, from this circumstance, to know what *were* the facts observed. Thus, Faujas de St. Fond, in his work on the volcanoes of central France, describes with every appearance of minute precision craters existing no where but in his own imagination. There is no greater fault (direct falsification of fact excepted) which can be committed by an observer.

(126.) When particular branches of science have acquired that degree of consistency and generality, which admits of an abstract statement of laws, and legitimate deductive reasoning, the principle of the division of labour tends to separate the province of the observer from that of the theorist. There is no accounting for the difference of minds or inclinations, which leads one man to observe with interest the developements of phenomena, another to speculate on their causes; but were it not for this happy disagreement, it may be doubted whether the higher sciences could ever have attained even their present degree of perfection. As laws acquire generality, the influence of individual observations becomes less, and a higher and higher degree of refinement in their performance, as well as a great multiplication in their number, becomes necessary to give them importance. In astronomy, for instance, the superior departments of theory are completely disjoined from the routine of practical observation.

(127.) To make a perfect observer, however, either in astronomy or in any other department of science, an extensive acquaintance is requisite, not only with the particular science to which his observations relate, but with every branch of knowledge which may enable him to appreciate and neutralize the effect of extraneous disturbing causes. Thus furnished, he will be prepared to seize on any of those minute indications, which (such is the subtlety of nature) often connect phenomena which seem quite remote from each other. He will have his eyes as it were opened, that they may be struck at once with any occurrence which, according to received theories, ought not to happen; for these are the facts which serve as clews to new discoveries. The deviation of the magnetic needle, by the influence of an electrified wire, must have happened a thousand times to a perceptible amount, under the eyes of persons engaged in galvanic experiments, with philosophical apparatus of all kinds standing around them; but it required the eye of a philosopher such as Oërsted to seize the indication, refer it to its origin, and thereby connect two great branches of science. The grand discovery of Malus of the polarization of light by reflection originated in his casual remark of the disappearance of one of the images of a window in the Luxembourg palace, one evening, when strongly illuminated by the setting sun, viewed through a doubly refracting prism.

(128.) To avail ourselves as far as possible of the advantages which a division of labour may afford for the collection of facts, by the industry and activity which the general diffusion of information, in the present age, brings into exercise, is an object of great importance. There is scarcely any well-informed person, who, if he has but the will, has not also the power to add something essential to the general stock of knowledge, if he will only observe regularly and methodically some particular class of facts which may most excite his attention, or which his situation may best enable him to study with effect. To instance one or two subjects, which *can* only be effectually improved by the united observations of great numbers widely dispersed: — Meteorology, one of the most complicated but important branches of science, is at the same time one in which any person who will attend to plain rules, and bestow the necessary degree of attention, may do effectual service. What benefits has not Geology reaped from the activity of industrious individuals, who, setting aside all theoretical views, have been content to exercise the useful and highly entertaining occupation of collecting specimens from the countries which they visit? In short, there is no branch of science whatever in which, at least, if useful and sensible queries were distinctly proposed, an immense mass of valuable information might not be collected from those who, in their various lines of life, at home or abroad, stationary or in travel, would gladly avail themselves of opportunities of being useful. Nothing would tend better to attain this end than the circulation of printed skeleton forms, on various subjects, which should be so formed as, 1st, to ask distinct and pertinent questions, admitting of short and definite answers; 2dly, To call for exact numerical statement on all principal points; 3dly, To point out the attendant circumstances most likely to prove influential, and which ought to be observed; 4thly, To call for their transmission to a common centre.

CHAP. V.

OF THE CLASSIFICATION OF NATURAL OBJECTS AND PHENOMENA, AND OF NOMENCLATURE.

(129.) The number and variety of objects and relations which the observation of nature brings before us are so great as to distract the attention, unless assisted and methodized by such judicious distribution of them in classes as shall limit our view to a few at a time, or to groups so bound together by general resemblances that, for the immediate purpose for which we consider them, they may be regarded as individuals. Before we can enter into any thing which deserves to be called a general and systematic view of nature, it is necessary that we should possess an enumeration, if not complete, at least of considerable extent, of her materials and combinations; and that those which appear in any degree important should be distinguished by names which may not only tend to fix them in our recollection, but may constitute, as it were, nuclei or centres, about which information may collect into masses. The imposition of a name on any subject of contemplation, be it a material object, a phenomenon of nature, or a group of facts and relations, looked upon in a peculiar point of view, is an epoch in its history of great importance. It not only enables us readily to refer to it in conversation or writing, without circumlocution, but, what is of more consequence, it gives it a recognized existence in our own minds, as a matter for separate and peculiar consideration; places it on a list for examination; and renders it a head or title, under which information of various descriptions may be arranged; and, in consequence, fits it to perform the office of a connecting link between all the subjects to which such information may refer.

(130.) For these purposes, however, a temporary or provisional name, or one adapted for common parlance, may suffice. But when a very great multitude of objects come to be referred to one class, especially of such as do not offer very obvious and remarkable distinctions, a more systematic and regular nomenclature becomes necessary, in which the names shall recall the differences as well as the resemblances between the individuals of a class, and in which the direct relation between the name and the object shall materially assist the solution of the problem, "*given the one, to determine the other*." How necessary this may become, will be at once seen, when we consider the immense number of individual objects, or rather species, presented by almost every branch of science of any extent; which absolutely require to be distinguished by names. Thus, the botanist is conversant with from 80,000 to 100,000 species of plants; the entomologist with, perhaps, as many, of insects: the chemist has to register the properties of combinations, by twos, threes, fours, and upwards, in various doses, of upwards of fifty different elements, all distinguished from each other by essential differences; and of which though a great many thousands are known, by far the greater part have never yet been formed, although hundreds of new ones are coming to light, in perpetual succession, as the science advances; all of which are to be named as they arise. The objects of astronomy are, literally, as numerous as the stars of heaven; and although not more than one or two thousand require to be expressed by distinct names, yet the number, respecting which particular information is required, is not less than a hundred times that amount; and all these must be registered in lists, (so as to be at once referred to, and so that none shall escape,) if not by actual names, at least by some equivalent means.

(131.) Nomenclature, then, is, in itself, undoubtedly an important part of science, as it prevents our being lost in a wilderness of particulars, and involved in inextricable confusion. Happily, in those great branches of science where the objects of classification are most numerous, and the necessity for a clear and convenient nomenclature most pressing, no very great difficulty in its establishment is felt. The very multitude of the objects themselves affords the power of grouping them in subordinate classes, sufficiently well defined to admit of names, and these again into others, whose names may become attached to, or compounded with, the former, till at length the particular species is identified. The facility with which the botanist, the entomologist, or the chemist, refers by name to any individual object in his science shows what may be accomplished in this way when characters are themselves distinct. In other branches, however, considerable difficulty is experienced. This arises mostly where the species to be distinguished are separated from each other chiefly by difference in degree, of certain qualities common to all, and where the degrees shade into each other insensibly. Perhaps such subjects can hardly be considered ripe for systematic nomenclature; and that the attempt to apply it ought only to be partial, embracing such groups and parcels of individuals as agree in characters evidently natural and generic, and leaving the remainder under trivial or provisional denominations, till they shall be better known, and capable of being scientifically grouped.

(132.) Indeed, nomenclature, in a systematic point of view, is as much, perhaps more, a consequence than a cause of extended knowledge. Any one may give an arbitrary name to a thing, merely to be able to talk of it; but, to give a name which shall at once refer it to a place in a system, we must know its properties; and we must *have* a system, large enough, and regular enough, to receive it in a place which belongs to it, and to no other. It appears, therefore, doubtful whether it is desirable, for the essential purposes of science, that extreme refinement in systematic nomenclature should be insisted on. Were science perfect, indeed, systems of classification might be agreed on, which should assign to every object in nature a place in some class, to which it more remarkably and pre-eminently belonged than to any other, and under which it might acquire a name, never afterwards subject to change. But, so long as this is not the case, and new relations are daily discovered, we must be very cautious how we insist strongly on the establishment and extension of classes which have in them any thing artificial, as a basis of a rigid nomenclature; and especially how we mistake the means for the end, and sacrifice convenience and distinctness to a rage for arrangement. Every nomenclature dependent on artificial classifications is necessarily subject to fluctuations; and hardly any thing can counterbalance the evil of disturbing well-established names, which have once acquired a general circulation. In nature, one and the same object makes a part of an infinite number of different systems, — an individual in an infinite number of groups, some of greater, some of less importance, according to the different points of view in which they may be considered. Hence, as many different systems of nomenclature may be imagined as there can be discovered different heads of classification, while yet it is highly desirable that each object should be universally spoken of under one name, *if possible*. Consequently, in all subjects where comprehensive heads of classification do not prominently offer themselves, all nomenclature must be a balance of difficulties, and a good, short, *unmeaning* name, which has once obtained a footing in usage, is preferable to almost any other.

(133.) There is no science in which the evils resulting from a rage for nomenclature have been felt to such an extent as in mineralogy. The number of simple minerals actually recognised by mineralogists does not exceed a few hundreds, yet there is scarcely one which has not four or five names in different books. The consequence is most unhappy. No name is suffered to endure long enough to take root; and every new writer on this interesting science begins, as a matter of course, by making a *tabula rasa* of all former nomenclature, and proposing a new one in its place. The climax has at length been put to this most inconvenient and bewildering state of things by the appearance of a system supported by extraordinary merit in other respects, and therefore carrying the highest authority, in which names which had acquired universal circulation, and had hitherto maintained their ground in the midst of the general confusion, and even worked their way into common language, as denotive of *species* too definite to admit of mistake, are actually rendered *generic*, and extended to whole groups, comprising objects agreeing in nothing but the arbitrary heads of a classification from which the most important natural relations are professedly and purposely rejected.*

(134.) The classifications by which science is advanced, however, are widely different from those which serve as bases for artificial systems of nomenclature. They cross and intersect one another, as it were, in every possible way, and have for their very aim to interweave all the objects of nature in a close and compact web of mutual relations and dependence. As soon, then, as any resemblance or analogy, any point of agreement whatever, is perceived between any two or more things,—be they what they will, whether objects, or phenomena, or laws,—they im-

* In the system alluded to, the name of quartz is assigned to iolite and obsidian; that of mica to plumbago, chlorite, and uranite; sulphur, to orpiment and realgar, &c. See Mohs's System of Mineralogy, translated by Haidinger.

mediately and *ipso facto* constitute themselves into a group or class, which may become enlarged to any extent by the accession of such new objects, phenomena, or laws, agreeing in the same point, as may come to be subsequently ascertained. It is thus that the materials of the world become grouped in natural families, such as chemistry furnishes examples of, in its various groups of acids, alkalies, sulphurets, &c.; or botany, in its euphorbiaceæ, umbelliferæ, &c. It is thus, too, that phenomena assume their places under general points of resemblance; as, in optics, those which refer themselves to the class of periodic colours, double refraction, &c.; and that resemblances themselves become traced, which it is the business of induction to generalize and include in abstract propositions.

(135.) But every class formed on a positive resemblance of characters, or on a distinct analogy, draws with it the consideration of a negative class, in which that resemblance either does not subsist at all, or the contrary takes place; and again, there are classes in which a given quality is possessed by the different individuals in a descending scale of intensity. Now, it is of consequence to distinguish between cases in which there is a real opposition of quality, or a mere diminution of intensity, in some quality susceptible of degrees, till it becomes imperceptible. For example, between transparency and opacity there would at first sight appear a direct opposition; but, on nearer consideration, when we consider the gradations by which transparency diminishes in natural substances, we shall see reason to admit that the latter quality, instead of being the *opposite* of the former, is only its *extreme lowest degree.* Again, in the arrangement of natural objects under the head of weight or specific gravity, the scale extends through all nature, and we know of no natural body in which the opposite of gravity, or positive *levity*, subsists. On the other hand, the opposite electricities; the north and south magnetic polarities; the alkaline and acid qualities of chemical agents; the positive and negative rotations impressed by plates of rock crystal on the planes of polarization of the rays of light, and many other cases, exemplify not merely a negation, but an active opposition of quality. Both these modes of classification have their peculiar importance in the inductive process: the one, as affording an opportunity of tracing a relation between phenomena by the observation of a correspondence in their scales of intensity; the other, by that of contrast, as we shall show more at large in the next section.

(136.) There is a very wide distinction, too, to be taken between such classes as turn upon a single head of resemblance among individuals otherwise very different, and such as bind together in natural groups, by a great variety of analogies, objects which yet differ in many remarkable particulars. For example: if we make colourless transparency a head of classification, the list of the class will comprise objects differing most widely in their nature, such as water, air, diamond, spirit of wine, glass, &c. On the other hand, the chemical families of alkalies, metals, &c. are instances of groups of the other kind; which, with properties in many respects different, still agree in a general resemblance of several others, which at once decides us in considering them as having a natural relation. In the former cases, our ingenuity is exercised to determine what can be the cause of their resemblance, in the latter, of their difference; the former belong to the province of inductive generalization, and afford the most instructive cases for the investigation of causes; the latter appertain to the more secret recesses of nature; the very existence of such families being in itself one of the great and complicated phenomena of the universe, which we cannot hope to unriddle without an intimate and extensive acquaintance with the highest laws.*

* The following passage, from Lindley's Synopsis of the British Flora, characterises justly the respective merits, in a philosophical point of view, of natural and artificial systems of classification in general, though limited in its expression to his own immediate science: — " After all that has been effected, or is likely to be accomplished hereafter, there will always be more difficulty in acquiring a knowledge of the natural system of botany than of the Linnæan. The latter skims only the surface of things, and leaves the student in the fancied possession of a sort of information which it is easy enough to obtain, but which is of little value when acquired: the former requires a minute investigation of every part and every property known to exist in plants; but when understood has conveyed to the mind a store of real information, of the utmost use to man in every station of life. Whatever the difficulties may be of becoming acquainted with plants according to this method, they are inseparable from botany, which cannot be usefully studied without encountering them." Schiller has some beautiful lines on this, entitled " Menschliches Wissen " (or Human Knowledge); Gedichte, vol. i. p. 72. Leipzig, 1800.

CHAP. VI.

OF THE FIRST STAGE OF INDUCTION.—THE DISCOVERY OF PROXIMATE CAUSES, AND LAWS OF THE LOWEST DEGREE OF GENERALITY, AND THEIR VERIFICATION.

(137.) THE first thing that a philosophic mind considers, when any new phenomenon presents itself, is its *explanation*, or reference to an immediate producing cause. If that cannot be ascertained, the next is to *generalize* the phenomenon, and include it, with others analogous to it, in the expression of some law, in the hope that its consideration, in a more advanced state of knowledge, may lead to the discovery of an adequate proximate cause.

(138.) Experience having shown us the manner in which one phenomenon depends on another in a great variety of cases, we find ourselves provided, as science extends, with a continually increasing stock of such antecedent phenomena, or causes (meaning at present merely proximate causes), competent, under different modifications, to the production of a great multitude of effects, besides those which originally led to a knowledge of them. To such causes Newton has applied the term *veræ causæ;* that is, causes recognized as having a real existence in nature, and not being mere hypotheses or figments of the mind. To exemplify the distinction:—The phenomenon of shells found in rocks, at a great height above the sea, has been attributed to several causes. By some it has been ascribed to a plastic virtue in the soil; by some, to fermentation; by some, to the influence of the celestial bodies; by some, to the casual passage of pilgrims with their scallops; by some, to birds feeding on shell-fish; and by all modern geologists, with one consent, to the life and death of real mollusca at the bottom of the sea, and a subsequent alteration of the relative level of the land and sea. Of these, the plastic virtue and celestial influence belong to the class of figments of fancy. Casual transport by pilgrims is a real cause, and might account for a few shells here and there dropped on frequented passes, but is not extensive enough for the purpose of explanation. Fermentation, generally, is a real cause, so far as that there *is such a thing;* but it is not a real cause of the production of a shell in a rock, since no such thing was ever witnessed as one of its effects, and rocks and stones do not ferment. On the other hand, for a shell-fish dying at the bottom of the sea to leave his shell in the mud, where it becomes silted over and imbedded, happens daily; and the elevation of the bottom of the sea to become dry land has really been witnessed so often, and on such a scale, as to qualify it for a *vera causa* available in sound philosophy.

(139.) To take another instance, likewise drawn from the same deservedly popular science: — The fact of a great change in the general climate of large tracts of the globe, if not of the whole earth, and of a diminution of general temperature, having been recognised by geologists, from their examination of the remains of animals and vegetables of former ages enclosed in the strata, various causes for such diminution of temperature have been assigned. Some consider the whole globe as having gradually cooled from absolute fusion; some regard the immensely superior activity of former volcanoes, and consequent more copious communication of internal heat to the surface, in former ages, as the cause. Neither of these can be regarded as real causes in the sense here intended; for we do not *know* that the globe has so cooled from fusion, nor are we sure that such supposed greater activity of former than of present volcanoes really did exist. A cause, possessing the essential requisites of a *vera causa*, has, however, been brought forward* in the varying influence of the distribution of land

* Lyell's Principles of Geology, vol. i. Fourrier, Mém. de l'Acad. des Sciences, tom. vii. p. 592. "L'établissement et le progrès des sociétés humaines, l'actions des forces naturelles, peuvent changer notablement, et dans de vastes contrées, l'état de la surface du sol, la distribution des eaux, et les grands mouvemens de l'air. De tels effets sont propres à faire varier, dans le cours de plusieurs siècles, le dégré de la chaleur moyenne; car les expressions analytiques comprennent des coefficiens qui se rapportent à l'état superficiel, et qui influent beaucoup sur la valeur de la température." In this enumeration, by M. Fourrier, of causes which may vary the general relation of the surface of extensive continents to heat, it is but justice to Mr. Lyell to observe, that the gradual shifting of the *places* of the continents themselves on the surface of the globe, by the abrading action of the sea on the one hand, and the elevating agency of subterranean forces on the other, does not expressly occur and cannot be fairly included in the general sense of the passage, which confines itself to the consideration of such changes as may take place on the existing surface of the land.

and sea over the surface of the globe: a change of such distribution, in the lapse of ages, by the degradation of the old continents, and the elevation of new, being a demonstrated fact; and the influence of such a change on the climates of particular regions, if not of the whole globe, being a perfectly fair conclusion, from what we know of continental, insular, and oceanic climates by actual observation. Here, then, we have, at least, a cause on which a philosopher may consent to reason; though, whether the changes actually going on are such as to warrant the whole extent of the conclusion, or are even taking place in the right direction, may be considered as undecided till the matter has been more thoroughly examined.

(140.) To this we may add another, which has likewise the essential characters of a *vera causa*, in the astronomical *fact* of the actual slow diminution of the eccentricity of the earth's orbit round the sun; and which, as a general one, affecting the *mean temperature of the whole globe*, and as one of which the effect is both inevitable, and susceptible, to a certain degree, of exact estimation, deserves consideration. It is evident that the *mean* temperature of the whole surface of the globe, in so far as it is maintained by the action of the sun at a higher degree than it would have were the sun extinguished, must depend on the mean quantity of the sun's rays which it receives, or, which comes to the same thing, on the *total* quantity received in a given invariable time: and the length of the year being unchangeable in all the fluctuations of the planetary system, it follows, that the total *annual* amount of solar radiation will determine, *cæteris paribus*, the general climate of the earth. Now, it is not difficult to show that this amount is inversely proportional to the minor axis of the ellipse described by the earth about the sun, regarded as slowly variable; and that, therefore, the major axis remaining, as we know it to be, constant, and the orbit being actually in a state of approach to a circle, and, consequently, the minor axis being on the *increase*, the mean annual amount of solar radiation received by the whole earth must be actually on the *decrease*. We have here, therefore, an evident real cause, of sufficient universality, and acting *in the right direction*, to account for the phenomenon. Its adequacy is another consideration.*

(141.) Whenever, therefore, any phenomenon presents itself for explanation, we naturally seek, in the first instance, to refer it to some one or other of those real causes which experience has shown to exist, and to be efficacious in producing similar phenomena. In this attempt our probability of success will, of course, mainly depend, 1st, On the number and variety of causes experience has placed at our disposal; 2dly, On our habit of applying them to the explanation of natural phenomena; and, 3dly, On the number of analogous phenomena we can collect, which have either been explained, or which admit of explanation by some one or other of those causes, and the closeness of their analogy with that in question.

* The reader will find this subject further developed in a paper lately communicated to the Geological Society.

(142.) Here, then, we see the great importance of possessing a stock of analogous instances or phenomena which class themselves with that under consideration, the explanation of one among which may naturally be expected to lead to that of all the rest. If the analogy of two phenomena be very close and striking, while, at the same time, the cause of one is very obvious, it becomes scarcely possible to refuse to admit the action of an analogous cause in the other, though not so obvious in itself For instance, when we see a stone whirled round in a sling, describing a circular orbit round the hand, keeping the string stretched, and flying away the moment it breaks, we never hesitate to regard it as retained in its orbit by the tension of the string, that is, by *a force* directed to the centre; for we feel that we do really exert such a force. We have here *the direct perception* of the cause. When, therefore, we see a great body like the moon circulating round the earth and not flying off, we cannot help believing it to be prevented from so doing, not indeed by a material tie, but by that which operates in the other case through the intermedium of the string, — a *force* directed constantly to the centre. It is thus that we are continually acquiring a knowledge of the existence of causes acting under circumstances of such concealment as effectually to prevent their direct discovery.

(143.) In general we must observe that motion, wherever produced or changed, invariably points out the existence of *force* as its cause; and thus the forces of nature become known and measured by the motions they produce. Thus, the *force* cf magnetism becomes known by the deviation produced by iron in a compass needle, or by a needle leaping up to a magnet held over it, as certainly as by that adhesion to it, when in contact and at rest, which requires force to break the connection; and thus the currents produced in the surface of a quantity of quicksilver, electrified under a conducting fluid, have pointed out the existence and direction of forces of enormous intensity developed by the electric circuit, of which we should not otherwise have had the least suspicion.*

(144.) But when the cause of a phenomenon neither presents itself obviously on the consideration of the phenomenon itself, nor is as it were forced on our attention by a case of strong analogy, such as above described, we have then no resource but in a deliberate assemblage of all the parallel instances we can muster; that is, to the formation of a class of facts, having the phenomenon in question for a head of classification; and to a search among the individuals of this class for some other common points of agreement, among which the cause will of necessity be found. But if more than one cause should appear, we must then endeavour to find, or, if we cannot find, to *produce, new facts,* in which each of these in succession shall be wanting, while yet they agree in the general point in question. Here we find the use of what Bacon terms "*crucial instances,*" which are phenomena brought forward to decide between two causes, each having the same analogies in its favour. And here, too, we perceive the utility

* Phil. Trans. 1824.

of *experiment* as distinguished from mere passive observation. We make an experiment of the crucial kind when we form combinations, and put in action causes from which some particular one shall be deliberately excluded, and some other purposely admitted; and by the agreement or disagreement of the resulting phenomena with those of the class under examination, we decide our judgment.

(145.) When we would lay down general rules for guiding and facilitating our search, among a great mass of assembled facts, for their common cause, we must have regard to the characters of that relation which we intend by cause and effect. Now, these are, —

1st, Invariable connection, and, in particular, invariable antecedence of the cause and consequence of the effect, unless prevented by some counteracting cause. But it must be observed, that, in a great number of natural phenomena, the effect is produced gradually, while the cause often goes on increasing in intensity; so that the antecedence of the one and consequence of the other becomes difficult to trace, though it really exists. On the other hand, the effect often follows the cause so instantaneously, that the interval cannot be perceived. In consequence of this, it is sometimes difficult to decide, of two phenomena constantly accompanying one another, which is cause or which effect.

2d, Invariable negation of the effect with absence of the cause, unless some other cause be capable of producing the same effect.

3d, Increase or diminution of the effect, with

the increased or diminished intensity of the cause, in cases which admit of increase and diminution.

4th, Proportionality of the effect to its cause in all cases of *direct unimpeded* action.

5th, Reversal of the effect with that of the cause.

(146.) From these characters we are led to the following observations, which may be considered as so many propositions readily applicable to particular cases, or rules of philosophizing: we conclude, 1st, That if in our group of facts there be one in which any assigned peculiarity, or attendant circumstance, is wanting or opposite, such peculiarity cannot be the cause we seek.

(147.) 2d, That any circumstance in which all the facts without exception agree, *may* be the cause in question, or, if not, at least a collateral effect of the same cause: if there be but one such point of agreement, this possibility becomes a certainty; and, on the other hand, if there be more than one, they may be concurrent causes.

(148.) 3d, That we are not to deny the existence of a cause in favour of which we have a unanimous agreement of strong analogies, though it may not be apparent how such a cause can produce the effect, or even though it may be difficult to conceive its existence under the circumstances of the case; in such cases we should rather appeal to experience when possible, than decide *à priori* against the cause, and try whether it cannot be made apparent.

(149.) For instance: seeing the sun vividly luminous, every analogy leads us to conclude it intensely hot. How heat can produce light, we know not;

and how such a heat can be maintained, we can form no conception. Yet we are not, therefore, entitled to deny the inference.

(150.) 4th, That contrary or opposing facts are equally instructive for the discovery of causes with favourable ones.

(151.) For instance: when air is confined with moistened iron filings in a close vessel over water its bulk is diminished, by a certain portion of it being abstracted and combining with the iron, producing *rust.* And, if the remainder be examined, it is found that it will *not* support flame or animal life. This contrary fact shows that the cause of the support of flame and animal life is to be looked for in that part of the air which the iron abstracts, and which rusts it.

(152.) 5th, That causes will very frequently become obvious, by a mere arrangement of our facts in the order of intensity in which some peculiar quality subsists; though not of necessity, because counteracting or modifying causes may be at the same time in action.

(153.) For example: sound consists in impulses communicated to our ears by the air. If a series of impulses of equal force be communicated to it at equal intervals of time, at first in slow succession, and by degrees more and more rapidly, we hear at first a rattling noise, then a low murmur, and then a hum, which by degrees acquires the character of a musical note, rising higher and higher in acuteness, till its pitch becomes too high for the ear to follow. And from this correspondence between the pitch of the note and the rapidity of succession of the impulse, we conclude that our sensation of the different pitches of musical notes originates in the different rapidities with which their impulses are communicated to our ears.

(154.) 6th, That such counteracting or modifying causes may subsist unperceived, and annul the effects of the cause we seek, in instances which, but for their action, would have come into our class of favourable facts; and that, therefore, exceptions may often be made to disappear by removing or allowing for such counteracting causes. This remark becomes of the greatest importance, when (as is often the case) a single striking exception stands out, as it were, against an otherwise unanimous array of facts in favour of a certain cause.

(155.) Thus, in chemistry, the *alkaline* quality of the alkaline and earthy bases is found to be due to the presence of oxygen combined with one or other of a peculiar set of metals. Ammonia is, however, a violent outstanding exception, such as here alluded to, being a compound of azote and hydrogen: but there are almost certain indications that this exception is not a real one, but assumes that appearance in consequence of some modifying cause not understood.

(156.) 7th, If we can either find produced by nature, or produce designedly for ourselves, two instances which agree *exactly* in all but one particular, and differ in that one, its influence in producing the phenomenon, if it have any, *must* thereby be rendered sensible. If that particular be present in one instance and wanting altogether in the other, the production or non-production of the phenomenon will decide whether it be or be not the only cause: still more evidently, if it be present *contrariwise* in the two cases, and the effect be thereby reversed. But if its total presence or absence only produces a change in the *degree* or intensity of the phenomenon, we can then only conclude that it acts as a concurrent cause or condition with some other to be sought elsewhere. In nature, it is comparatively rare to find instances pointedly differing in one circumstance and agreeing in every other; but when we call experiment to our aid, it is easy to produce them; and this is, in fact, the grand application of *experiments of enquiry* in physical researches. They become more valuable, and their results clearer, in proportion as they possess this quality (of agreeing exactly in all their circumstances but one), since the question put to nature becomes thereby more pointed, and its answer more decisive.

(157.) 8th, If we cannot obtain a complete negative or opposition of the circumstance whose influence we would ascertain, we must endeavour to find cases where it varies considerably in degree. If *this* cannot be done, we may perhaps be able to weaken or exalt its influence by the introduction of some fresh circumstance, which, abstractedly considered, seems *likely* to produce this effect, and thus obtain indirect evidence of its influence. But then we are always to remember, that the evidence so obtained *is* indirect, and that the new circumstance introduced *may* have a direct influence of its own, or may exercise a modifying one on some *other* circumstance.

(158.) 9th, Complicated phenomena, in which several causes concurring, opposing, or quite independent of each other, operate at once, so as to produce a compound effect, may be simplified by subducting the effect of all the known causes, as well as the nature of the case permits, either by deductive reasoning or by appeal to experience, and thus leaving, as it were, a *residual phenomenon* to be explained. It is by this process, in fact, that science, in its present advanced state, is chiefly promoted. Most of the phenomena which nature presents are very complicated; and when the effects of all known causes are estimated with exactness, and subducted, the residual facts are constantly appearing in the form of phenomena altogether new, and leading to the most important conclusions.

(159.) For example: the return of the comet predicted by professor Encke, a great many times in succession, and the general good agreement of its calculated with its observed place during any one of its periods of visibility, would lead us to say that its gravitation towards the sun and planets is the sole and sufficient cause of all the phenomena of its orbitual motion; but when the effect of this cause is strictly calculated and subducted from the observed motion, there is found to remain behind a *residual phenomenon*, which would never have been otherwise ascertained to exist, which is a small anticipation of the time of its reappearances or a diminution of its periodic time, which cannot be accounted for by gravity, and whose cause is therefore to be enquired into. Such an anticipation would be caused by the resistance of a medium disseminated through the celestial regions; and as there are other good reasons for believing this to be a *vera causa*, it has therefore been ascribed to such a resistance.

(160.) This 9th observation is of such importance in science, that we shall exemplify it by another instance or two. M. Arago, having suspended a magnetic needle by a silk thread, and set it in vibration, observed, that it came much sooner to a state of rest when suspended over a plate of copper, than when no such plate was beneath it. Now, in both cases there were two *veræ causæ* why it *should* come at length to rest, viz. the resistance of the air, which opposes, and at length destroys, all motions performed in it; and the want of perfect mobility in the silk thread. But the effect of these causes being exactly known by the observation made in the absence of the copper, and being thus allowed for and subducted, a *residual* phenomenon appeared, in the fact that a retarding influence was exerted by the copper itself; and this fact, once ascertained, speedily led to the knowledge of an entirely new and unexpected class of relations. To add one more instance. If it be true (as M. Fourrier considers it demonstrated to be) that the celestial regions have a temperature independent of the sun, not greatly inferior to that at which quicksilver congeals, and much superior to some degrees of cold which have been artificially produced, two causes suggest themselves: one is that assigned by the author above mentioned; the radiation of the stars; another may be proposed in the ether or elastic medium mentioned in the last section, which the phenomena of light and the resistance of comets give us reason to believe fills all space, and which, in analogy to all the elastic media known, may be supposed to possess a temperature and a specific heat of its own, which it is capable of communicating to bodies surrounded by it. Now, if we consider that the heat radiated by the sun follows the same proportion as its light, and regard it as reasonable to admit with respect to stellar heat what holds good of solar; the effect of stellar radiation in maintaining a temperature in space should be as much inferior to that of the radiation of the sun as the light of a moonless midnight is to that of an equatorial noon; that is to say, almost inconceivably smaller. Allowing, then, the full effect for this cause, there would still remain a great residuum due to the presence of the ether.

(161.) Many of the new elements of chemistry have been detected in the investigation of *residual phenomena*. Thus, Arfwedson discovered lithia by perceiving an *excess of weight* in the sulphate produced from a small portion of what he considered as magnesia present in a mineral he had analysed. It is on this principle, too, that the *small concentrated residues of great operations* in the arts are almost sure to be the lurking places of new chemical ingredients: witness iodine, brome, selenium, and the new metals accompanying platina in the experiments of Wollaston and Tennant. It was a happy thought of Glauber to examine what every body else threw away.

(162.) Finally, we have to observe, that the detection of a *possible* cause, by the comparison of assembled cases, *must* lead to one of two things: either, 1st, The detection of a real cause, and of its manner of acting, so as to furnish a complete explanation of the facts; or, 2dly, The establishment of an abstract law of nature, pointing out two phenomena of a general kind as invariably connected; and asserting, that where one is, there the other will always be found. Such invariable connection is itself a phenomenon of a higher order than any particular fact; and when many such are discovered, we may again proceed to classify, combine, and examine them, with a view to the detection of *their* causes, or the discovery of still more general laws, and so on without end.

(163.) Let us now exemplify this inductive search for a cause by one general example: suppose *dew* were the phenomenon proposed, whose cause we would know. In the first place, we must separate dew from rain and the moisture of fogs, and limit the application of the term to what is really meant, which is, the spontaneous appearance of moisture on substances exposed in the open air when no rain or *visible* wet is falling. Now, here we have analogous phenomena in the moisture which bedews a cold metal or stone when we breathe upon it; that which appears on a glass of water fresh from the well in hot weather; that which appears on the *inside* of windows when sudden rain or hail chills the external air; that which runs down our walls when, after a long frost, a warm moist thaw comes on: all these instances agree in one point (Rule 2. § 147.), the coldness of the object dewed, in comparison with the air in contact with it.

(164.) But, in the case of the night dew, is this a *real cause*—is it a fact that the object dewed *is* colder than the air? Certainly not, one would at first be inclined to say; for what is to *make* it so? But the analogies are cogent and unanimous; and, therefore, (pursuant to Rule 3. § 148.) we are not to discard their indications; and, besides, the experiment is easy: we have only to lay a thermometer in contact with the dewed substance, and hang one at a little distance above it out of reach of its influence. The experiment has been therefore made; the question has been asked, and the answer has been invariably in the *affirmative*. Whenever an object contracts dew, *it is* colder than the air. Here, then, we have *an invariable concomitant* circumstance: but is this chill an effect of dew, or its cause? That dews are accompanied with a chill is a common remark; but vulgar prejudice would make the cold the *effect* rather than the cause. We must, therefore, collect more facts, or, which comes to the same thing, vary the circumstances; since every instance in which the circumstances differ is a fresh fact; and, especially, we must note the contrary or negative cases (Rule 4. § 150.), *i. e.* where no dew is produced.

(165.) Now, 1st, no dew is produced on the surface of *polished metals*, but it *is* very copiously on glass, both exposed with their faces upwards, and in some cases the under side of a horizontal plate of glass is also dewed; which last circumstance (by Rule 1. § 146.) excludes the *fall* of moisture from the sky in an invisible form, which would naturally suggest itself as a cause. In the cases of polished metal and polished glass, the contrast shows evidently that the *substance* has much to do with the phenomenon; therefore, let the substance *alone* be diversified as much as possible, by exposing polished surfaces of various kinds. This done, *a scale of intensity* becomes obvious (Rule 5. § 152.). Those polished substances are found to be most strongly dewed which conduct heat worst; while those which conduct well resist dew most effectually. Here we encounter a *law* of the first degree of generality. But, if we expose rough surfaces, instead of polished, we sometimes find this law interfered with (Rule 5. § 152.). Thus, roughened iron, especially if painted over or blackened, becomes dewed sooner than varnished paper: the kind of *surface* therefore has a great influence. Expose, then, the *same* material in very diversified states as to surface (Rule 7. § 156.), and another scale of intensity becomes at once apparent; those *surfaces* which *part with their heat* most readily by radiation are found to contract dew most copiously: and thus we have detected another law of the same generality with the former, by a comparison of two classes of facts, one relating to dew, the other to the radiation of heat from surfaces. Again, the influence ascertained to exist of *substance* and *surface* leads us to consider that of *texture:* and here, again, we are presented on trial with remarkable differences, and with a third *scale of intensity*, pointing out substances of a close firm texture, such as stones, metals, &c. as unfavourable, but those of a loose one, as cloth, wool, velvet, eiderdown, cotton, &c. as eminently favourable, to the contraction of dew: and these are precisely those which are best adapted for clothing, or for impeding the free passage of heat from the skin into the air, so as to allow their outer surfaces to be very cold while they remain warm within.

(166.) Lastly, among the negative instances, (§ 150.) it is observed, that dew is never copiously deposited in situations much screened from the open sky, and not at all in *a cloudy night;* but if the clouds withdraw, even for a few minutes, and leave a clear opening, a deposition of dew presently begins, and goes on increasing. Here, then, a cause is distinctly pointed out by its antecedence to the effect in question (§ 145.). A clear view of the cloudless sky, then, is an essential condition, or, which comes to the same thing, clouds or surrounding objects act as *opposing causes.* This is so much the case, that dew formed in clear intervals will often even evaporate again when the sky becomes thickly overcast (Rule 4. § 150.).

(167.) When we now come to assemble these partial inductions so as to raise from them a general conclusion, we consider, 1st, That all the conclusions we have come to have a reference to that first general fact — the cooling of the exposed surface of the body dewed below the temperature of the air. Those surfaces which part with their heat outwards most readily, and have it supplied from within most slowly, will, of course, become coldest if there be an opportunity for their heat to escape, and not be restored to them from without. Now, a clear sky affords such an opportunity. It is a law well known to those who are conversant with the nature of heat, that heat is constantly escaping from *all bodies* in rays, or by *radiation,* but is as constantly restored to them by the similar radiation of others surrounding them. Clouds and surrounding objects therefore act as opposing causes by replacing the whole or a great part of the heat so radiated away, which can escape effectually, without being replaced, only through openings into infinite space. Thus, at length, we arrive at the general proximate cause of dew, in the cooling of the dewed surface by radiation faster than its heat can be restored to it, by communication with the ground, or by counter-radiation; so as to become colder than the air, and thereby to cause a condensation of its moisture.

(168.) We have purposely selected this theory of dew, first developed by the late Dr. Wells, as one of the most beautiful specimens we can call to mind of inductive experimental enquiry lying within a moderate compass. It is not possible in so brief a space to do it justice; but we earnestly recommend his work* (a short and very entertaining one) for perusal to the student of natural philosophy, as a model with which he will do well to become familiar.

(169.) In the analysis above given, the formation of dew is referred to two more general phenomena; the radiation of heat, and the condensation of invisible vapour by cold. The cause of the former is a much higher enquiry, and may be said, indeed, to be totally unknown; that of the latter actually forms a most important branch of physical enquiry. In such a case, when we reason upwards till we reach an ultimate fact, we regard a phenomenon as fully explained; as we consider the branch of a tree to

* Wells on Dew

terminate when traced to its insertion in the trunk, or a twig to its junction with the branch; or rather, as a rivulet retains its importance and its name till lost in some larger tributary, or in the main river which delivers it into the ocean. This, however, always supposes that, on a reconsideration of the case, we see clearly how the admission of such a fact, with all its attendant laws, will perfectly account for *every particular*—as well those which, in the different stages of the induction, have led us to a knowledge of it, as those which we had neglected, or considered less minutely than the rest. But, had we no previous knowledge of the radiation of heat this same induction would have made it known to us, and, duly considered, might have led to the knowledge of many of its laws.

(170.) In the study of nature, we must not, therefore, be scrupulous as to *how* we reach to a knowledge of such general facts: provided only we verify them carefully when once detected, we must be content to seize them wherever they are to be found. And this brings us to consider the *verification* of inductions.

(171.) If, in our induction, every individual case has actually been present to our minds. we are sure that it will find itself duly *represented* in our final conclusion: but this is impossible for such cases as were *unknown* to us and hardly ever happens even with all the known cases; for such is the tendency of the human mind to speculation, that on the least idea of an analogy between a few phenomena, it leaps forward, as it were, to a cause or law, to the temporary neglect of all the rest; so that, in fact, almost all our principal inductions must be regarded as a series of ascents and descents, and of conclusions from a few cases, verified by trial on many.

(172.) Whenever, therefore, we think we have been led by induction to the knowledge of the proximate cause of a phenomenon or of a law of nature, our next business is to examine deliberately and *seriatim* all the cases we have collected of its occurrence, in order to satisfy ourselves that they are explicable by our cause, or fairly included in the expression of our law: and in case any exception occurs, it must be carefully noted and set aside for re-examination at a more advanced period, when, possibly, the cause of exception may appear, and the exception itself, by allowing for the effect of that cause, be brought over to the side of our induction; but should exceptions prove numerous and various in their features, our faith in the conclusion will be proportionally shaken, and at all events its importance lessened by the destruction of its universality.

(173.) In the conduct of this verification, we are to consider whether the cause or law to which we are conducted be one already known and recognised as a more general one, whose nature is well understood, and of which the phenomenon in question is but one more case in addition to those already known, or whether it be one less general, less known, or altogether new. In the latter case, our verification will suffice, if it merely shows that all the cases considered are plainly cases in point. But in the former, the process of verification is of a much more severe and definite kind. We must trace the action of our cause with distinctness and precision, as modi-tied by all the circumstances of each case; we must estimate its effects, and show that nothing unexplained remains behind; at least, in so far as the presence of unknown modifying causes is not concerned.

(174.) Now, this is precisely the sort of process in which *residual phenomena* (such as spoken of in art. 158.) may be expected to occur. If our induction be really a valid and a comprehensive one, *whatever* remains unexplained in the comparison of its conclusion with particular cases, under all their circumstances, *is* such a phenomenon, and comes in its turn to be a subject of inductive reasoning to discover its cause or laws. It is thus that we may be said to witness facts with the eyes of reason; and it is thus that we are continually attaining a knowledge of new phenomena and new laws which lie beneath the surface of things, and give rise to the creation of fresh branches of science more and more remote from common observation.

(175.) Physical astronomy affords numerous and splendid instances of this. The law, for example, which asserts that the planets are retained in their orbits about the sun, and satellites about their primaries, by an attractive force, decreasing as the square of the distances increases, comes to be verified in each particular case by deducing from it the exact motions which, under the circumstances, ought to take place, and comparing them with fact. This comparison, while it verifies in general the existence of the law of gravitation as supposed, and its adequacy to explain all the principal motions of every body in the system, yet leaves some small deviations in those of the planets, and some very considerable ones in that of the moon and other satellites, still unaccounted for; residual phenomena, which still remain to be traced up to causes. By further examining these, their causes have at length been ascertained, and found to consist in the mutual actions of the planets on each other, and the disturbing influence of the sun on the motions of the satellites.

(176.) But a law of nature has not that degree of generality which fits it for a stepping-stone to greater inductions, unless it be *universal* in its application. We cannot rely on its enabling us to extend our views beyond the circle of instances from which it was obtained, unless we have already had experience of its power to do so; unless it actually *has* enabled us before trial to say what will take place in cases analogous to those originally contemplated; unless, in short, we have studiously placed ourselves in the situation of its antagonists, and even perversely endeavoured to find exceptions to it without success. It is in the precise proportion that a law once obtained endures this extreme severity of trial, that its value and importance are to be estimated; and our next step in the verification of an induction must therefore consist in *extending* its application to cases not originally contemplated: in studiously varying the circumstances under which our causes act, with a view to ascertain whether their effect is general; and in pushing the application of our laws to extreme cases.

(177.) For example, a fair induction from a

great number of facts led Galileo to conclude that the accelerating power of gravity is the same on all sorts of bodies, and on great and small masses indifferently; and this he exemplified by letting bodies of very different natures and weights fall at the same instant from a high tower, when it was observed that they struck the ground at the same moment, abating a certain trifling difference, due, as he justly believed it to be, to the greater proportional resistance of the air to light than to heavy bodies. The experiment could not, at that time, be fairly tried with extremely light substances, such as cork, feathers, cotton, &c. because of the great resistance experienced by these in their fall; no means being then known of removing this cause of disturbance. It was not, therefore, till after the invention of the air-pump that this law could be put to the severe test of an extreme case. A guinea and a downy feather were let drop at once from the upper part of a tall exhausted glass, and struck the bottom at the same moment. Let any one make the trial *in the air*, and he will perceive the force of an *extreme case*.

(178.) In the verification of a law whose expression is *quantitative*, not only must its generality be established by the trial of it in as various circumstances as possible, but every such trial must be one of precise measurement. And in such cases the means taken for subjecting it to trial ought to be so devised as to repeat and multiply a great number of times any deviation (if any exist); so that, let it be ever so small, it shall at last become sensible.

(179.) For instance, let the law to be verified

be, that *the gravity of every material body is in the direct proportion of its mass*, which is only another mode of expressing Galileo's law above mentioned. The time of falling from any moderate height cannot be measured with precision enough for our purpose: but if it can be repeated a very great multitude of times *without any loss or gain* in the intervals, and the whole amount of the times of fall so repeated measured by a clock; and if at the same time the resistance of the air can be rendered *exactly alike* for all the bodies tried, we have here Galileo's trial in a much more refined state; and it is evident that almost unlimited exactness may be obtained. Now, all this Newton accomplished by the simple and elegant contrivance of enclosing in a hollow pendulum the same weights of a great number of substances the most different that could be found in all respects, as gold, glass, wood, water, wheat, &c.*, and ascertaining the time required for the pendulum so charged to make a great number of oscillations; in each of which it is clear the weights had to fall, and be raised again successively, without loss of time, through the same *identical* spaces. Thus any difference, however inconsiderable, that might exist in the time of one such fall and rise would be multiplied and accumulated till they became sensible. And none having been discovered by so delicate a process in any case, the law was considered verified both in respect of generality and exactness. This, however, is nothing to the verifications afforded by astronomical phenomena, where the deviations, if any, accumulate for thousands of years instead of a few hours.

* Principia, book iii. prop. 6.

(180.) The surest and best characteristic of a well-founded and extensive induction, however, is when verifications of it spring up, as it were, spontaneously, into notice, from quarters where they might be least expected, or even among instances of that very kind which were at first considered hostile to them. Evidence of this kind is irresistible, and compels assent with a weight which scarcely any other possesses. To give an example: M. Mitscherlich had announced a law to this effect — *that* the chemical elements of which all bodies consist are susceptible of being classified in distinct groups, which he termed *isomorphous* groups; and *that* these groups are so related that when similar combinations are formed of individuals belonging to two, three, or more of them, such combinations will crystallize in the same geometrical forms. To this curious and important law there appeared a remarkable exception. According to professor Mitscherlich, the arsenic and phosphoric acids *are* similar combinations coming under the meaning of his law, and their combinations with soda and water, forming the salts known to chemists under the names of arseniate and phosphate of soda, ought, if the law were general, to crystallize in identical shapes. The fact, however, was understood to be otherwise. But lately, Mr. Clarke, a British chemist, having examined the two salts attentively, ascertained the fact that their compositions deviate essentially from that similarity which M. Mitscherlich's law requires; and that, therefore, the exception in question disappears. This was something: but, pursuing the subject further, the same ingenious enquirer happily succeeded in producing a *new* phosphate of soda, differing from that generally known in containing a different proportion of water, and agreeing in composition exactly with the arseniate. The crystals of this new salt, when examined, were found by him to be precisely identical in form with those of the arseniate: thus verifying, in a most striking and totally unexpected manner, the law in question, or, as it is called, the law of isomorphism.

(181.) Unexpected and peculiarly striking confirmations of inductive laws frequently occur in the form of residual phenomena, in the course of investigations of a widely different nature from those which gave rise to the inductions themselves. A very elegant example may be cited in the unexpected confirmation of the law of the developement of heat in elastic fluids by compression, which is afforded by the phenomena of sound. The enquiry into the cause of sound had led to conclusions respecting its mode of propagation, from which its velocity in the air could be precisely calculated. The calculations were performed; but, when compared with fact, though the agreement was quite sufficient to show the general correctness of the cause and mode of propagation assigned, *yet* the *whole* velocity could not be shown to arise from this theory. There was still a *residual* velocity to be accounted for, which placed dynamical philosophers for a long time in a great dilemma. At length La Place struck on the happy idea, that this might arise from the *heat* developed in the act of that condensation which necessarily takes place at every vibration by which sound is conveyed. The matter was subjected to exact calculation, and the result was at once the complete explanation of the residual phenomenon, and a striking confirmation of the general law of the developement of heat by compression, under circumstances beyond artificial imitation.

(182.) In extending our inductions to cases not originally contemplated, there is one step which always strikes the mind with peculiar force, and with such a sensation of novelty and surprise, as often gives it a weight beyond its due philosophic value. It is the transition from the little to the great, and *vice versâ*, but especially the former. It is so beautiful to see, for instance, an experiment performed in a watch-glass, or before a blowpipe, succeed, in a great manufactory, on many tons of matter, or, in the bosom of a volcano, upon millions of cubic fathoms of lava, that we almost forget that these great masses are made up of watch-glassfuls, and blowpipe-beads. We see the enormous intervals between the stars and planets of the heavens, which afford room for innumerable processes to be carried on, for light and heat to circulate, and for curious and complicated motions to go forward among them: we look more attentively, and we see sidereal systems, probably not less vast and complicated than our own, crowded apparently into a small space (from the effect of their distance from us), and forming groups resembling bodies of a substantial appearance, having form and outline: yet we recoil with incredulous surprise when we are asked *why* we cannot conceive the atoms of a grain of sand to be as remote from each other (proportionally to their sizes) as the stars of the firmament; and why there may not be going on, in that little microcosm, processes as complicated and wonderful as those of the great world around us. Yet the student who makes any progress in natural philosophy will encounter numberless cases in which this transfer of ideas from the one extreme of magnitude to the other will be called for: he will find, for instance, the phenomena of the propagation of winds referred to the same laws which regulâte the propagation of motions through the smallest masses of air; those of lightning assimilated to the mere communication of an electric spark, and those of earthquakes to the tremors of a stretched wire: in short, he must lay his account to finding the distinction of great and little altogether annihilated in nature: and it is well for man that such is the case, and that the same laws, which he can discover and verify in his own circumscribed sphere of power, should prove available to him when he comes to apply them on the greatest scale; since it is thus only that he is enabled to become an exciting cause in operations of any considerable magnitude, and to vindicate his importance in creation.

(183.) But the business of induction does not end here: its final result must be followed out into all its consequences, and applied to all those cases which seem even remotely to bear upon the subject of enquiry. Every new addition to our stock of causes becomes a means of fresh attack with new vantage ground upon all those unexplained parts of former phenomena which have resisted previous efforts. It can hardly be pressed forcibly enough on the attention of the student of nature, that there is scarcely any natural phenomenon which can be fully and completely explained in all its circumstances, without a union of several, perhaps of all, the sciences. The great phenomena of astronomy, indeed, may be considered exceptions; but this is merely because their scale is so vast that one only of the most widely extending forces of nature takes the lead, and all those agents whose sphere of action is limited to narrower bounds, and which determine the production of phenomena nearer at hand, are thrown into the back ground, and become merged and lost in comparative insignificance. But in the more intimate phenomena which surround us it is far otherwise. Into what a complication of different branches of science are we not led by the consideration of such a phenomenon as rain, for instance, or flame, or a thousand others, which are constantly going on before our eyes? Hence, it is hardly possible to arrive at the knowledge of a law of any degree of generality in any branch of science, but it immediately furnishes us with a means of extending our knowledge of innumerable others, the most remote from the point we set out from; so that, when once embarked in any physical research, it is impossible for any one to predict where it may ultimately lead him.

(184.) This remark rather belongs to the inverse or *deductive* process, by which we pursue laws into their remote consequences. But it is very important to observe, that the successful process of scientific enquiry demands continually the alternate use of both the *inductive* and *deductive* method. The path by which we rise to knowledge must be made smooth and beaten in its lower steps, and often ascended and descended, before we can scale our way to any eminence, much less climb to the summit. The achievement is too great for a single effort; stations must be established, and communications kept open with all below. To quit metaphor; there is nothing so instructive, or so likely to lead to the acquisition of general views, as this pursuit of the consequences of a law once arrived at into every subject where it may seem likely to have an influence. The discovery of a new law of nature, a new ultimate fact, or one that even temporarily puts on that appearance, is like the discovery of a new element in chemistry. Thus, selenium was hardly discovered by Berzelius in the vitriol works of Fahlun, when it presently made its appearance in the sublimates of Stromboli, and the rare and curious products of the Hungarian mines. And thus it is with every new law, or general fact. It is hardly announced before its traces are found every where, and every one is astonished at its having so long remained concealed. And hence it happens that unexpected lights are shed at length over parts of science that had been abandoned in despair, and given over to hopeless obscurity.

(185.) The verification of *quantitative* laws has been already spoken of (178.); but their importance in physical science is so very great, inasmuch as they alone afford a handle to strict mathematical deductive application, that something ought to be said of the nature of the inductions by which they are to be arrived at. In their simplest or least general stages (of which alone we speak at present) they usually express some numerical relation between two quantities dependent on each other, either as collateral effects of a common cause, or as the amount of its effect under given numerical circumstances or *data*. For example, the law of refraction before noticed (§ 22.) expresses, by a very simple relation, the amount of angular deviation of a ray of light from its course, when the *angle* at which it is inclined to the refracting surface is known, viz. that the *sine* of the angle which the incident ray makes with a perpendicular to the surface is always to that of the angle made by the refracted ray with the same perpendicular, in a constant proportion, so long as the refracting substance is the same. To arrive inductively at laws of this kind, where one quantity *depends* on or *varies with* another, all that is required is a series of careful and exact measures in every different state of the *datum* and *quæsitum*. Here, however, the mathematical form of the law being of the highest importance, the greatest attention must be given to the *extreme cases* as well as to all those points where the one quantity changes rapidly with a small change of the other.* The results must be set down in a table in which the *datum* gradually increases in magnitude from the lowest to the highest limit of which it is sus-

* A very curious instance of the pursuit of a law completely empirical into an extreme case is to be found in Newton's rule for the dilatation of his coloured rings seen between glasses at great obliquities. Optics, book ii. part i. obs. 7.

ceptible. It will depend then entirely on our habit of treating mathematical subjects, how far we may be able to include such a table in the distinct statement of a mathematical law. The discovery of such laws is often remarkably facilitated by the contemplation of a class of phenomena to be noticed further on, under the head of Collective Instances, (see § 194.) in which the nature of the mathematical expression in which the law sought is comprehended, is pointed out by the figure of some curve brought under inspection by a proper mode of experimenting.

(186.) After all, unless our induction embraces a series of cases which absolutely include the whole scale of variation of which the quantities in question admit, the mathematical expression so obtained cannot be depended upon as the true one, and if the scale actually embraced be small, the extension of laws so derived to extreme cases will in all probability be exceedingly fallacious. For example, air is an elastic fluid, and as such, if enclosed in a confined space and squeezed, its bulk diminishes: now, from a great number of trials made in cases where the air has been compressed into a half, a third, &c. even as far as a fiftieth of its bulk, or less, it has been concluded that "the density of air is proportional to the compressing force," or the bulk it occupies *inversely* as that force; and when the air is rarefied by taking off part of its natural pressure, the same is found to be the case, within very extensive limits. Yet it is impossible that this should be, strictly or mathematically speaking, the true law; for, if it were so, there could be no limit to the condensation of air, while yet we have the strongest analogies to show that long before it had reached any very enormous pitch the air would be reduced into a liquid, and even, perhaps, if pressed yet more violently, into a solid form.

(187.) Laws thus derived, by the direct process of including in mathematical formulæ the results of a greater or less number of measurements, are called "empirical laws." A good example of such a law is that given by Dr. Young (Phil. Trans. 1826,) for the decrement of life, or the law of mortality. Empirical laws in this state are evidently *unverified inductions*, and are to be received and reasoned on with the utmost reserve. No confidence can ever be placed in them beyond the limits of the data from which they are derived; and even within those limits they require a special and severe scrutiny to examine *how nearly* they do represent the observed facts; that is to say, whether, in the comparison of their results with the observed quantities, the differences are such as may fairly be attributed to error of observation. When so carefully examined, they become, however, most valuable; and frequently, when afterwards verified theoretically by a deductive process (as will be explained in our next chapter), turn out to be rigorous laws of nature, and afford the noblest and most convincing supports of which theories themselves are susceptible. The finest instances of this kind are the great laws of the planetary motions deduced by Kepler, entirely from a comparison of observations with each other, with no assistance from theory. These laws, viz. that the planets move in ellipses round the sun; that each describes about the sun's centre equal areas in equal times; and that in the orbits of different planets the squares of the periodical times are proportional to the cubes of the distances; were the results of inconceivable labour of calculation and comparison: but they amply repaid the labour bestowed on them, by affording afterwards the most conclusive and unanswerable proofs of the Newtonian system. On the other hand, when empirical laws are unduly relied on beyond the limits of the observations from which they were deduced, there is no more fertile source of fatal mistakes. The formulæ which have been empirically deduced for the elasticity of steam (till very recently), and those for the resistance of fluids, and other similar subjects, have almost invariably failed to support the theoretical structures which have been erected on them.

(188.) It is a remarkable and happy fact, that the shortest and most direct of all inductions should be that which has led at once, and almost by a single step, to the highest of all natural laws,—we mean those of motion and force. Nothing can be more simple, precise, and general, than the enunciation of these laws; and, as we have once before observed, their application to particular facts in the descending or deductive method is limited by nothing but the limited extent of our mathematics. It would seem, then, that dynamical science were taken thenceforward out of the pale of induction, and transformed into a matter of absolute *à priori* reasoning, as much as geometry; and so it would be, were our mathematics perfect, and all the *data* known. Unhappily, the first is so far from being the case, that in many of the most interesting branches of dynamical enquiry they leave us completelv at a loss. In what relates to the motions of fluids, for instance, this is severely felt. We can include our problems, it is true, in algebraical equations, and we can demonstrate that they *contain* the solutions; but the equations themselves are so intractable, and present such insuperable difficulties, that they often leave us quite as much in the dark as before. But even were these difficulties overcome, recourse to experience must still be had, to establish the *data* on which particular applications are to depend; and although mathematical analysis affords very powerful means of *representing* in general terms the data of any proposed case, and *afterwards*, by comparison of its results with fact, determining *what* those data must be to explain the observed phenomena, still, in any mode of considering the matter, an appeal to experience in every particular instance of application is unavoidable, even when the general principles are regarded as sufficiently established without it. Now, in all such cases of difficulty we must recur to our inductive processes, and regard the branches of dynamical science where this takes place as purely experimental. By this we gain an immense advantage, viz. that in all those points of them where the abstract dynamical principles *do* afford distinct conclusions, we obtain verifications for our inductions of the highest and finest possible kind. When we work our way up inductively to one of these results, we cannot help feeling the strongest assurance of the validity of the induction.

(189.) The necessity of this appeal to experiment in every thing relating to the motions of fluids on the large scale has long been felt. Newton himself, who laid the first foundations of hydrodynamical science (so this branch of dynamics is called), distinctly perceived it, and set the example of laborious and exact experiments on their resistance to motion, and other particulars. Venturi, Bernoulli, and many others, have applied the method of experiment to the motions of fluids in pipes and canals; and recently the brothers Weber have published an elaborate and excellent experimental enquiry into the phenomena of waves. One of the greatest and most successful attempts, however, to bring an important, and till then very obscure, branch of dynamical enquiry back to the dominion of experiment, has been made by Chladni and Savart in the case of sound and vibratory motion in general; and it is greatly to be wished that the example may be followed in many others hardly less abstruse and impracticable when theoretically treated. In such cases the inductive and deductive methods of enquiry may be said to go hand in hand, the one verifying the conclusions deduced by the other; and the combination of experiment and theory, which may thus be brought to bear in such cases, forms an engine of discovery infinitely more powerful than either taken separately. This state of any department of science is perhaps of all others the most interesting, and that which promises the most to research.

(190.) It can hardly be expected that we should terminate this division of our subject without some mention of the "prerogatives of instances" of Bacon, by which he understands characteristic phenomena, selected from the great miscellaneous mass of facts which occur in nature, and which, by their number, indistinctness, and complication, tend rather to confuse than to direct the mind in its search for causes and general heads of induction. Phenomena so selected on account of some peculiarly forcible way in which they strike the reason, and impress us with a kind of sense of causation, or a particular aptitude for generalization, he considers, and justly, as holding a kind of prerogative dignity, and claiming our first and especial attention in physical enquiries.

(191.) We have already observed that, in forming inductions, it will most commonly happen that we are led to our conclusions by the especial force of some two or three strongly impressive facts, rather than by affording the whole mass of cases a regular consideration; and hence the need of cautious verification. Indeed, so strong is this propensity of the human mind, that there is hardly a more common thing than to find persons ready to assign a cause for every thing they see, and, in so doing, to join things the most incongruous, by analogies the most fanciful. This being the case, it is evidently of great importance that these first ready impulses of the mind should be made on the contemplation of the cases most likely to lead to good inductions. The misfortune, however, is, in natural philosophy, that the choice does not rest with us. We must take the instances as nature presents them. Even if we are furnished with a list of them in tabular order, we must understand and compare them with each other, before we can tell which *are* the instances thus deservedly entitled to the highest consideration. And, after all, after much labour in vain, and groping in the dark, accident or casual observation will present a case which strikes us at once with a full insight into a subject, before we can even have time to determine to what class its *prerogative* belongs. For example, the laws of crystallography were obscure, and its causes still more so, till Haüy fortunately dropped a beautiful crystal of calcareous spar on a stone pavement, and broke it. In piecing together the fragments, he observed their facets not to correspond with those of the crystal in its entire state, but to belong to another form; and, following out the hint offered by a "*glaring instance*" thus casually obtruded on his notice, he discovered the beautiful laws of the cleavage, and the primitive forms of minerals.

(192.) It has always appeared to us, we must confess, that the help which the classification of instances, under their different titles of prerogative, affords to inductions, however just such classification may be in itself, is yet more apparent than real. The force of the instance must be felt in the mind, before it can be referred to its place in the system; and, before it can be either referred or appretiated, it must be known; and when it *is* appretiated, we are ready enough to interweave it in our web of induction, without greatly troubling ourselves with enquiring whence it derives the weight we acknowledge it to have in our decisions. However, since much importance is usually attached to this part of

Bacon's work, we shall here give a few examples to illustrate the nature of some of his principal cases. One, of what he calls "glaring instances," has just been mentioned. In these, the *nature*, or cause enquired into, (which in this case is the cause of the assumption of a peculiar external form, or the internal *structure* of a crystal,) "stands naked and alone, and this in an eminent manner, or in the highest degree of its power." No doubt, such instances as these are highly instructive; but the difficulty in physics is to find such, not to perceive their force when found.

(193.) The contrary of glaring are "clandestine instances," where "the nature sought is exhibited in its weakest and most imperfect state." Of this, Bacon himself has given an admirable example in the cohesion of fluids, as a *clandestine instance* of the "*nature* or quality of consistence, or solidity." Yet here, again, the same acute discrimination which enabled Bacon to perceive the analogy which connects fluids with solids, through the common property of cohesive attraction, would, at the same time, have enabled him to draw from it, if properly supported, every consequence necessary to forming just notions of the cohesive force; nor does its reference to the class of clandestine instances at all assist in bringing forward and maturing the final results. When, however, the final result is obtained,—when our induction is complete, and we would verify it,—this class of instances is of great use, being, in fact, frequently no other than that of *extreme cases*, such as we have already spoken of (in § 177.); which, by placing our conclusions, as, it were, in violent circumstances, try their temper, and bring their vigour to the test.

(194.) "Collective instances," in Bacon's classification, are no other than general facts, or laws of some degree of generality, and are themselves the results of induction. But there is a species of collective instance which Bacon does not seem to have contemplated, of a peculiarly instructive character; and that is, where particular cases are offered to our observation in such numbers at once as to make the induction of their law a matter of ocular inspection. For example, the parabolic form assumed by a jet of water spouted from a round hole, is a *collective instance* of the velocities and directions of the motions of all the particles which compose it *seen at once*, and which thus leads us, without trouble, to recognize the law of the motion of a projectile. Again, the beautiful figures exhibited by sand strewed on regular plates of glass or metal set in vibration, are *collective instances* of an infinite number of points which remain at rest while the remainder of the plate vibrates; and in consequence afford us, as it were, a sight of the law which regulates their arrangement and sequence throughout the whole surface. The beautifully coloured lemniscates seen around the optic axes of crystals exposed to polarized light afford a superb example of the same kind, pointing at once to the general mathematical expression of the law which regulates their production.* Of such collective instances as these, it is easy to see the importance, and its reason. They lead us to a general law by an induction which

* See Phil. Trans. 1819.

offers itself spontaneously, and thus furnish advanced points in our enquiries; and when we start from these, already "a thousand steps are lost."

(195.) A fine example of a collective instance is that of the system of Jupiter or Saturn with its satellites. We have here, in miniature, and seen at one view, a system similar to that of the planets about the sun; of which, from the circumstance of our being involved in it, and unfavourably situated for seeing it otherwise than in detail, we are incapacitated from forming a general idea but by slow progressive efforts of reason. Accordingly, the contemplation of the *circumjovial planets* (as they were called) most materially assisted in securing the admission of the Copernican system.

(196.) Of "Crucial instances" we have also already spoken, as affording the readiest and securest means of eliminating extraneous causes, and deciding between rival hypotheses. Owing to the disposition of the mind to form hypotheses, and to prejudge cases, it constantly happens that, among all the possible suppositions which may occur, two or three principal ones occupy us, to the exclusion of the rest; or it may be that, if we have been less precipitate, out of a great multitude rejected for obvious inapplicability to some one or other case, two or three of better claims remain for decision; and this such instances enable us to do. One of the instances cited by Bacon in illustration of his crucial class is very remarkable, being neither more nor less than the proposal of a direct experiment to determine whether the tendency of heavy bodies downwards is a result of some peculiar mechanism in themselves, or of the attraction of the earth "by the corporeal mass thereof, as by a collection of bodies of the same nature." If it be so, he says, "it will follow that the nearer all bodies approach to the earth, the stronger and with the greater force and velocity they will tend to it; but the farther they are, the weaker and slower:" and his experiment consists in comparing the effect of a spring and a weight in keeping up the motions of two "clocks," regulated together, and removed alternately to the tops of high buildings and into the deepest mines. By *clocks* he could not have meant pendulum clocks, which were not then known, (the first made in England was in 1662,) but *fly*-clocks, so that the comparison, though too coarse, was not contrary to sound mechanical principles. In short, its principle was the comparison of the effect of a spring with that of a weight, in producing certain motions in certain times, on heights and in mines. Now, this is the very same thing that has really been done in the recent experiments of professors Airy and Whewell in Dolcoath mine: a pendulum (a weight moved by gravity) has been compared with a chronometer balance, moved and regulated by a spring. In his 37th aphorism, Bacon also speaks of gravity as an incorporeal power, acting at a distance, and *requiring time for its transmission*; a consideration which occurred at a later period to Laplace, in one of his most delicate investigations.

(197.) A well chosen and strongly marked crucial instance is, sometimes, of the highest importance; when two theories, which run parallel to each other (as is sometimes the case) in their explanation of great classes of phenomena, at length come to be placed at issue upon a single fact. A beautiful instance of this will be cited in the next section. We may add to the examples above given of such instances, that of the application of chemical tests, which are almost universally crucial experiments.

(198.) Bacon's "travelling instances" are those in which the *nature* or quality under investigation "travels," or varies in degree; and thus (according to § 152.) afford an indication of a cause by a gradation of intensity in the effect. One of his instances is very happy, being that of "paper, which is white when dry, but proves less so when wet, and comes nearer to the state of transparency upon the exclusion of the air, and admission of water." In reading this, and many other instances in the Novum Organum, one would almost suppose (had it been written) that its author had taken them from Newton's Optics.

(199.) The travelling instances, as well as what Bacon terms "frontier instances," are cases in which we are enabled to trace that general law which seems to pervade all nature — the law, as it is termed, of continuity, and which is expressed in the well known sentence, "Natura non agit per saltum." The pursuit of this law into cases where its application is not at first sight obvious, has proved a fertile source of physical discovery, and led us to the knowledge of an analogy and intimate connection of phenomena between which at first we should never have expected to find any.

(200.) For example, the transparency of gold leaf,

which permits a bluish-green light to pass through it, is a frontier instance between the transparency of pellucid bodies and the opacity of metals, and it prevents a breach of the law of continuity between transparent and opake bodies, by exhibiting a body of the class generally regarded the most opake in nature, as still possessed of some slight degree of transparency. It thus proves that the quality of opacity is not a *contrary* or *antagonist* quality to that of transparency, but only its extreme lowest degree.

CHAP. VII.

OF THE HIGHER DEGREES OF INDUCTIVE GENERALIZATION, AND OF THE FORMATION AND VERIFICATION OF THEORIES.

(201.) As particular inductions and laws of the first degree of generality are obtained from the consideration of individual facts, so Theories result from a consideration of these laws, and of the proximate causes brought into view in the previous process, regarded all together as constituting a new set of phenomena, the creatures of reason rather than of sense, and each representing under general language innumerable particular facts. In raising these higher inductions, therefore, more scope is given to the exercise of pure reason than in slowly groping out our first results. The mind is more disencumbered of matter, and moves as it were in its own element. What is now before it, it perceives more intimately, and less through the medium of sense, or at least not in the same manner as when actually at work on the immediate objects of sense. But it must not be therefore supposed that, in the formation of theories, we are abandoned to the unrestrained exercise of imagination, or at liberty to lay down arbitrary principles, or assume the existence of mere fanciful causes. The liberty of speculation which we possess in the domains of theory is not like the wild licence of the slave broke loose from his fetters, but rather like that of the freeman who has learned the lessons of self-restraint in the school of just subordination. The ultimate objects we pursue in the highest theories are the same as those of the lowest inductions; and the means by which we can most securely attain them bear a close analogy to those which we have found successful in such inferior cases.

(202.) The immediate object we propose to ourselves in physical theories is the analysis of phenomena, and the knowledge of the hidden processes of nature in their production, so far as they can be traced by us. An important part of this knowledge consists in a discovery of the actual structure or mechanism of the universe and its parts, through which, and by which, those processes are executed; and of the agents which are concerned in their performance. Now, the mechanism of nature is for the most part either on too large or too small a scale to be immediately cognizable by our senses; and her agents in like manner elude direct observation, and become known to us only by their effects. It is in vain therefore that we desire to become witnesses to the processes carried on with such means, and to be admitted into the secret recesses and laboratories where they are effected. Microscopes have been constructed which magnify more than a thousand times in *linear* dimension, so that the smallest visible grain of sand may be enlarged to the appearance of one a thousand million times more bulky; yet the only impression we receive by viewing it through such a magnifier is, that it reminds us of some vast fragment of a rock, while the intimate structure on which depend its colour, its hardness, and its chemical properties, remains still concealed: we do not seem to have made even an approach to a closer analysis of it by any such scrutiny.

(203.) On the other hand, the mechanism of the great system of which our planet forms a part escapes immediate observation by the immensity of its scale, nay, even by the slowness of its evolutions. The motion of the minute hand of a watch can hardly be perceived without the closest attention, and that of the hour hand not at all. But what are these, in respect of the impression of slowness they produce in our minds, compared with a revolving movement which takes a whole year, or twelve, thirty, or eighty years to complete, as is the case with the planets in their revolutions round the sun. Yet no sooner do we come to reflect on the linear dimensions of these orbs, (which however we do not *see*, nor can we measure them but by a long, circuitous and difficult process,) than we are lost in astonishment at the swiftness of the very motions which before seemed so slow.* The motion of the sails of a windmill offers (on a small scale) an illustrative case. At a distance the rotation seems slow and steady—but when we stand close to one of the sails in its sweep, we are surprised at the swiftness with which it rushes by us.

* "When we are told that Saturn moves in his orbit more than 22,000 miles an hour, we fancy the motion to be swift; but when we find that he is more than three hours moving his own diameter, we must then think it, as it really is, slow." Thirty Letters on various Subjects, by William Jackson, 1795.

(204.) Again, the agents employed by nature to act on material structures are invisible, and only to be traced by the effects they produce. Heat dilates matter with an irresistible force; but what heat *is*, remains yet a problem. A current of electricity passing along a wire moves a magnetized needle at a distance; but except from this effect we perceive no difference between the condition of the wire when it conveys and when it does not convey the stream: and we apply the terms current, or stream, to the electricity only because in some of its relations it reminds us of something we have observed in a stream of air or water. In like manner we see that the moon circulates about the earth; and because we believe it to be a solid mass, and have never seen one solid substance revolve round another within our reach to handle and examine unless retained by a force or united by a tie, we conclude that there *is* a force, and a mode of connection, between the moon and the earth; though, what that mode can be, we have no conception, nor can imagine *how* such a force can be exerted at a distance, and with empty space, or at most an invisible fluid, between. (See § 148.)

(205.) Yet are we not to despair, since we see regular and beautiful results brought about in human works by means which nobody would, at first sight, think could have any thing to do with them. A sheet of blank paper is placed upon a frame, and shoved forwards, and after winding its way successively over and under half a dozen rollers, and performing many other strange evolutions, comes out printed on both sides. And, after all, the acting cause in this process is nothing more than a few gallons of water boiled in an iron vessel, at a distance from the scene of operations. But *why* the water so boiled should be capable of producing the active energy which sets the whole apparatus in motion is, and will probably long remain, a secret to us.

(206.) This, however, does not at all prevent our having a very perfect comprehension of the whole subsequent process. We might frequent printing-houses, and form a theory of printing, and having worked our way up to the point where the mechanical action commenced (the boiler of the steam-engine), and verified it by taking to pieces, and putting together again, the train of wheels and the presses, and by sound theoretical examination of all the transfers of motion from one part to another; we should, at length, pronounce our theory good, and declare that we understood printing thoroughly. Nay, we might even go away and apply the principles of mechanism we had learned in this enquiry to other widely different purposes; construct other machines, and put them in motion by the same moving power, and all without arriving at any correct idea as to the ultimate source of the force employed. But, if we were inclined to theorize farther, we might do so; and it is easy to imagine how two theorists might form very different *hypotheses* as to the origin of the power which alternately raised and depressed the piston-rod of the engine. One, for example, might maintain that the boiler (whose contents we will suppose that neither theorist has been permitted to examine) was the den of some powerful unknown animal, and he would not be without plausible analogies in the warmth, the supply of fuel and water, the breathing noises, the smoke, and above all, the mechanical power exerted. He would say (not without a show of reason), that where there is a positive and wonderful effect, and many strong analogies, such as materials consumed, and all the usual signs of life maintained, we are not to deny the existence of animal life because we know no animal that consumes such food. Nay, he might observe with truth, that the fuel actually consists of the chemical ingredients which constitute the chief food of all animals, &c.; while, on the other hand, his brother theorist, who caught a glimpse of the fire, and detected the peculiar sounds of ebullition, might acquire a better notion of the case, and form a theory more in consonance with fact.

(207.) Now, nothing is more common in physics than to find two, or even many, *theories* maintained as to the origin of a natural phenomenon. For instance, in the case of heat itself, one considers it as a really existing material fluid, of such exceeding subtlety as to penetrate all bodies, and even to be capable of combining with them chemically; while another regards it as nothing but a rapid vibratory or rotatory motion in the ultimate particles of the bodies heated; and produces a singularly ingenious train of mechanical reasoning to show, that there is nothing contradictory to sound dynamical principles in such a doctrine. Thus, again, with light: one considers it as consisting in actual particles darted forth from luminous bodies, and acted upon in their progress by forces of extreme intensity residing in the substances on which they strike; another, in the vibratory motion of the particles of luminous bodies, communicated to a peculiar subtle and highly elastic ethereal medium, filling all space, and conveyed through it into our eyes, as sounds are to our ears, by the undulations of the air.

(208.) Now, are we to be deterred from framing hypotheses and constructing theories, because we meet with such dilemmas, and find ourselves frequently beyond our depth? Undoubtedly not. *Est quodam prodire tenus si non datur ultra.* Hypotheses, with respect to theories, are what presumed proximate causes are with respect to particular inductions: they afford us motives for searching into analogies; grounds of citation to bring before us all the cases which seem to bear upon them, for examination. A well imagined hypothesis, if it have been suggested by a fair inductive consideration of general laws, can hardly fail at least of enabling us to generalize a step farther, and group together several such laws under a more universal expression. But this is taking a very limited view of the value and importance of hypotheses: it may happen (and it has happened in the case of the undulatory doctrine of light) that such a weight of analogy and probability may become accumulated on the side of an hypothesis, that we are compelled to admit one of two things; either that it is an actual statement of what really passes in nature, or that the reality, whatever it be, must run so close a parallel with it, as to admit of some mode of expression common to both, at least in so far as the phenomena actually known are concerned. Now, this is a very great step, not only for its own sake, as leading us to a high point in philosophical speculation, but for its applications; because whatever conclusions we deduce from an hypothesis so supported must have at least a strong presumption in their favour: and we may be thus led to the trial of many curious experiments, and to the imagining of many useful and important contrivances, which we should never otherwise have thought of, and which, at all events, *if* verified in practice, are real additions to our stock of knowledge and to the arts of life.

(209.) In framing a theory which shall render a rational account of any natural phenomenon, we have *first* to consider the agents on which it depends, or the causes to which we regard it as ultimately referable. These agents are not to be arbitrarily assumed; they must be such as we have good inductive grounds to believe do exist in nature, and do perform a part in phenomena analogous to those we would render an account of; or such, whose presence in the actual case can be demonstrated by unequivocal signs. They must be *veræ causæ*, in short, which we can not only show to exist and to act, but the laws of whose action we can derive independently, by direct induction, from experiments purposely instituted; or at least make such suppositions respecting them as shall not be contrary to our experience, and which will remain to be verified by the coincidence of the conclusions we shall deduce from them, with facts. For example, in the theory of gravitation we suppose an agent,—*viz.* force, or mechanical power,—to act on *any* material body which is placed in the presence of *any* other, and to urge the two mutually towards each other. This is a *vera causa;* for heavy bodies (that is, all bodies, but some more, some less,) tend to, or endeavour to reach, the earth, and require the exertion of force to counteract this endeavour, or to keep them up. Now, that which opposes and neutralizes force *is* force. And again, a plumb-line, which, when allowed to hang freely, always hangs perpendicularly; is found to hang observably aside from the perpendicular when in the neighbourhood of a considerable mountain; thereby proving that a force is exerted upon it, which draws it towards the mountain. Moreover, since it is a fact that the moon does circulate about the earth, it must be drawn towards the earth by a force; for if there were no force acting upon it, it would go on in a straight line without turning aside to circulate in an orbit, and would, therefore, soon go away and be lost in space. This force, then, which we call the *force* of gravity, is a real cause.

(210.) We have next to consider the laws which regulate the action of these our primary agents; and these we can only arrive at in three ways: 1st, By inductive reasoning; that is, by examining all the cases in which we know them to be exercised, inferring, as well as circumstances will permit, its amount or intensity in each particular case, and then piecing together, as it were, these *disjecta membra*, generalizing from them, and so arriving at the laws desired; 2dly, By forming at once a bold hypothesis, particularizing the law, and trying the truth of it by following out its consequences and comparing them with facts; or, 3dly, By a process partaking of both these, and combining the advantages of both without their defects, viz. by assuming indeed the laws we would discover, but so generally expressed, that they shall include an unlimited variety of particular laws; — following out the consequences of this assumption, by the application of such general principles as the case admits; — comparing them in succession with all the particular cases within our knowledge; and, lastly, *on this comparison*, so modifying and restricting the general enunciation of our laws as to *make the results agree.*

(211.) All these three processes for the discovery of those general elementary laws on which the higher theories are grounded are applicable with different advantage in different circumstances. We might exemplify their successive application to the case of gravitation; but as this would rather lead into a disquisition too particular for the objects of this discourse, and carry us too much into the domain of technical mathematics, we shall content ourselves with remarking, that the method last mentioned is that which mathematicians (especially such as have a considerable command of those general modes of representing and reasoning on quantity, which constitute the higher analysis,) find the most universally applicable, and the most efficacious; and that it is applicable with especial advantage in cases where subordinate inductions of the kind described in the last section have already led to laws of a certain generality admitting of mathematical expression. Such a case, for instance, is the elliptic motion of a planet, which is a general proposition including the statement of an infinite number of particular *places*, in which the laws of its motion allow it to be some time or other found, and for which, of course, the law of force must be so assumed as to account.

(212.) With regard to the first process of the three above enumerated, it is in fact an induction of the kind described in § 185.; and all the remarks we there made on that kind of induction apply to it in this stage. The direct assumption of a particular hypothesis has been occasionally practised very successfully. As examples, we may mention Coulomb's and Poisson's theories of electricity and magnetism, in both which, phenomena of a very complicated and interesting nature are referred to the actions of attractive and repulsive forces, following a law similar in its expression to the law of gravitation. But the difficulty and labour, which, in the greater theories, always attends the pursuit of a fundamental law into its remote consequences, effectually precludes this method from being commonly resorted to as a means of discovery, unless we have some good reason, from analogy or otherwise, for believing that the attempt will prove successful, or have been first led by partial inductions to particular laws which naturally point it out for trial.

(213.) In this case the law assumes all the characters of a general phenomenon resulting from an induction of particulars, but not yet verified by comparison with *all* the particulars, nor extended to all that it is capable of including. (See § 171.) It is the verification of such inductions which constitutes theory in its largest sense, and which embraces an estimation of the influence of all such circumstances as may modify the effect of the cause whose laws of action we have arrived at and would verify. To return to our example: particular inductions drawn from the motions of the several planets about the sun, and of the satellites round their primaries, &c. having led us to the general conception of an attractive force exerted by every particle of matter in the universe on every other according to the law to which we attach the name of gravitation; when we would verify this induction, we must set out with assuming this law, considering the whole system as subjected to its influence and implicitly obeying it, and nothing interfering with its action; we then, for the first time, perceive a train of modifying circumstances which had not occurred to us when reasoning upwards from particulars to obtain the fundamental law; we perceive that *all the planets* must attract *each other*, must therefore draw each other out of the orbits which they would have if acted on only by the sun; and as this was never contemplated in the inductive process, its validity becomes a question, which can only be determined by ascertaining precisely how great a deviation this new class of mutual actions will produce. To do this is no easy task, or rather, it is the most difficult task which the genius of man has ever yet accomplished: still, it *has* been accomplished by the mere application of the general laws of dynamics; and the result (undoubtedly a most beautiful and satisfactory one) is, that all those observed deviations in the motions of our system which stood out as exceptions (§ 154.), or were noticed as residual phenomena and reserved for further enquiry (§ 158.), in that imperfect view of the subject which we got in the subordinate process by which we rose to our general conclusion, prove to be the immediate consequences of the above-mentioned mutual actions. As such, they are neither exceptions nor residual facts, but fulfilments of general rules, and essential features in the statement of the case, *without* which our induction would be invalid, and the law of gravitation positively untrue.

(214.) In the theory of gravitation. the law is all in all, applying itself at once to the materials, and directly producing the result (A). But in many other cases we have to consider not merely the laws which regulate the actions of our ultimate causes, but a system of mechanism, or a structure of parts, through the intervention of which their effects become sensible to us. Thus, in the delicate and curious electro-dynamic theory of Ampere, the mutual attraction or repulsion of two magnets is referred to a more universal phenomenon, the mutual action of electric currents, according to a certain fundamental law. But, in order to bring the case of a magnet within the range of this law, he is obliged to make a supposition of a peculiar structure or mechanism, which constitutes a body a magnet, viz. that around each particle of the body there shall be constantly circulating, in a certain stated direction, a small current of electric fluid.

(215.) This, we may say, is too complex; it is artificial, and cannot be granted: yet, if the admission of this or any other structure tenfold more artificial and complicated will enable any one to present in a general point of view a great number of particular facts, — to make them a part of one system, and enable us to reason from the known to the unknown, and actually to *predict facts before trial*, — we would ask, why should it *not* be granted? When we examine those instances of nature's workmanship which we can take to pieces and understand, we find them in the highest degree artificial in our own sense of the word. Take, for example, the structure of an eye, or of the skeleton of an animal, — what complexity and what artifice! In the one, a *pellucid muscle;* a lens formed with elliptical surfaces; a circular aperture capable of enlargement or contraction without loss of form. In the other, a framework of the most curious carpentry; in which occurs not a single straight line, nor any known geometrical curve, yet all evidently systematic, and constructed by rules which defy our research. Or examine a crystallized mineral, which we can in some measure dissect, and thus obtain direct evidence of an internal structure. Neither artifice nor complication are here wanting; and though it is easy to assert that these appearances are, after all, produced by something which would be very simple, if we did but know it, it is plain that the same might be *said* of a steam-engine executing the most complicated movements, previous to any investigation of its nature, or any knowledge of the source of its power.

(216.) In estimating, however, the value of a theory, we are not to look, *in the first instance*, to the question, whether it establishes satisfactorily, or not, a particular process or mechanism; for of this, after all, we can never obtain more than that indirect evidence which consists in its leading to the same results. What, in the actual state of science, is far more important for us to know, is whether our theory truly represent *all* the facts, and include *all* the laws, to which observation and induction lead. A theory which did this would, no doubt, go a great way to establish any hypothesis of mechanism or structure, which might form an essential part of it: but this is very far from being the case, except in a few limited instances; and, till it is so, to lay any great stress on hypotheses of the kind, except in as much as they serve as a scaffold for the erection of general laws, is to "quite mistake the scaffold for the pile." Regarded in this light, hypotheses have often an eminent use: and a facility in framing them, if attended with an equal facility in laying them aside when they have served their turn, is one of the most valuable qualities a philosopher can possess; while, on the other hand, a bigoted adherence to them, or indeed to peculiar views of any kind, in opposition to the tenor of facts as they arise, is the bane of all philosophy.

(217.) There is no doubt, however, that the safest course, when it can be followed, is to rise by inductions carried on among laws, as among facts, from law to law, perceiving, as we go on, how laws which we have looked upon as unconnected become particular cases, either one of the other, or all of one still more general, and, at length, blend altogether in the point of view from which we learn to regard them. An example will illustrate what we mean. It is a general law, that all hot bodies throw out or *radiate* heat in all directions, (by which we mean, not that heat is an actual substance darted out from hot bodies, but only that the laws of the transmission of heat to distant objects are similar to those which would regulate the distribution of particles thrown forth in all directions,) and that other colder bodies placed in their neighbourhood become hot, *as if* they received the heat so radiated. Again, all solid bodies which become heated in one part *conduct*, or diffuse, the heat from that part through their whole substance. Here we have two modes of communicating heat,— by radiation, and by conduction; and both these have their peculiar, and, to all appearance, very different laws. Now, let us bring a hot and a cold body (of the same substance) gradually nearer and nearer together,— as they approach, the heat will be communicated from the hot to the cold one by the *laws of radiation;* and from the nearer to the farther part of the colder one, as it gradually grows warm, by *those of conduction.* Let their distance be diminished till they just lightly touch. How does the heat *now* pass from one to the other? Doubtless, by radiation; for it may be proved, that in such a contact there is yet an interval. Let them then be *forced* together, and it will seem clear that it must now be by *conduction.* Yet their *interval* must diminish gradually, as the force by which they are pressed together increases, till they actually cohere, and form one. The law of continuity, then, of which we have before spoken (§ 199.), forbids us to suppose that the intimate nature of the process of communication is changed in this transition from light to violent contact, and from that to actual union. If so, we might ask, at what point does the change happen? Especially since it is also demonstrable, that the particles of the most solid body are not, really, in contact. *Therefore*, the laws of conduction and radiation have a mutual dependence, and the former are only extreme cases of the latter. If, then, we would rightly understand what passes, or what is the process of nature in the slow communication of heat through the substance of a solid, we must ground our enquiries upon what takes place at a distance, and then urge the laws to which we have arrived, up to their extreme case.

(218.) When two theories run parallel to each other, and each explains a great many facts in common with the other, any experiment which affords a crucial instance to decide between them, or by which one or other must fall, is of great importance. In thus verifying theories, since they are grounded on general laws, we may appeal, not merely to particular cases, but to whole classes of facts; and we therefore have a great range among the individuals of these for the selection of some particular effect which ought to take place oppositely in the event of one of the two suppositions at issue being right and the other wrong. A curious example is given by M. Fresnel, as decisive, in his mind, of the question between the two great opinions on the nature of light, which, since the time of Newton and Huyghens, have divided philosophers. (See § 207.) When two very clean glasses are laid one on the other, if they be not perfectly flat, but one or both in an almost imperceptible degree convex or prominent, beautiful and vivid colours will be seen between them; and if these be viewed through a red glass, their appearance will be that of alternate dark and bright stripes. These stripes are formed *between* the two surfaces in apparent contact, as any one may satisfy himself by using, instead of a flat *plate* of glass for the upper one, a triangular-shaped piece, called a prism, like a three-cornered stick, and looking through the inclined side of it next the eye, by which arrangement the reflection of light from the upper surface is prevented from intermixing with that from the surfaces in contact. Now, the coloured stripes thus produced are explicable on both theories, and are appealed to by both as strong confirmatory facts; but there is a difference in one circumstance according as one or the other theory is employed to explain them. In the case of the Huyghenian doctrine, the intervals between the bright stripes ought to appear *absolutely black;* in the other, *half bright,* when so viewed through a prism. This curious case of difference was tried as soon as the opposing consequences of the two theories were noted by M. Fresnel, and the result is stated by him to be decisive in favour of that theory which makes light to consist in the vibrations of an elastic medium.

(219.) Theories are best arrived at by the consideration of general laws; but most securely verified by comparing them with particular facts, because this serves as a verification of the whole train of induction, from the lowest term to the highest. But then, the comparison must be made with facts purposely selected so as to include every variety of case, not omitting extreme ones, and in sufficient number to afford every reasonable probability of detecting error. A single numerical coincidence in a final conclusion, however striking the coincidence or important the subject, is not sufficient. Newton's theory of sound, for example, leads to a numerical expression for the actual velocity of sound, differing but little from that afforded by the correct theory afterwards explained by La Grange, and (when certain considerations not contemplated by him are allowed for) agreeing with fact; yet this coincidence is no verification of Newton's view of the general subject of sound, which is defective in an essential point, as the great geometer last named has very satisfactorily shown. This example is sufficient to inspire caution in resting the verification of theories upon any thing but a very extensive comparison with a great mass of observed facts.

(220.) But, on the other hand, when a theory will bear the test of such extensive comparison, it matters little how it has been originally framed. However strange and, at first sight, inadmissible its postulates may appear, or however singular it may seem that such postulates should have been fixed upon,— if they only lead us, by legitimate reasonings, to conclusions in exact accordance with numerous observations purposely made under such a variety of circumstances as fairly to embrace the whole range of the phenomena which the theory is intended to account for, we cannot refuse to admit them; or if we still hesitate to regard them as demonstrated truths, we cannot, at least, object to receive them as temporary substitutes for such truths, until the latter shall become known. If they suffice to explain all the phenomena known, it becomes highly improbable that they will not explain more; and if all their conclusions we have tried have proved correct, it is probable that others yet untried will be found so too; so that in rejecting them altogether, we should reject all the discoveries to which they may lead.

(221.) In all theories which profess to give a true account of the process of nature in the production of any class of phenomena, by referring them to general laws, or to the action of general causes, through a train of modifying circumstances; before we can apply those laws, or trace the action of those causes in any assigned case, we require to know the circumstances: we must have data whereon to ground their application. Now, these can be learned only from observation; and it may be seem to be arguing in a vicious circle to have recourse to observation for any part of those theoretical conclusions, by whose comparison with fact the theory itself is to be tried. The consideration of an example will enable us to remove this difficulty. The most general law which has yet been discovered in chemistry is this, that all the elementary substances in nature are susceptible of entering into combination with each other only in fixed or *definite proportions* by weight, and not arbitrarily; so that when any two substances are put together with a view to unite them, if their weights are not in some certain determinate proportion, a complete combination will not take place, but some part of one or the other ingredient will remain over and above, and uncombined. Suppose, now, we have found a substance having all the outward characters of a homogeneous or unmixed body, but which, on analysis, we discover to consist of sulphur and lead in the proportion of 20 parts of the former to 130 of the latter ingredient; and we would know whether this is to be regarded as a verification of the law of definite proportions or an exception to it. The question is reduced to this, whether the proportion 20 to 130 be or be not *that* fixed and definite proportion, (or one of them, if there be more than one proportion possible,) in which, according to the law in question, sulphur and lead can combine; now, this can never be decided by merely looking at the law in all its generality. It is clear, that when particularized by restricting its expression to sulphur and lead, the law should state *what are* those particular fixed proportions in which these bodies can combine. That is to say, there must be certain data or numbers, by which these are distinguished from all other bodies in nature, and which require to be known before we can apply the general law to the particular case. To determine such data, observation must be consulted; and if we were to have recourse to that of the combination of the two substances in question with each other, no doubt there would be ground for the logical objection of a vicious circle: but this is not done; the determination of these numerical data is derived from experiments purposely made on a great variety of different combinations, among which that under consideration does not of necessity occur, and all these being found, independently of each other, to agree in giving the same results, they are therefore safely assumed as part of the system. Thus, the law of definite proportions, when applied to the actual state of nature, requires two separate statements, the one announcing the general law of combination, the other particularizing the numbers appropriate to the several elements of which natural bodies consist, or the data of nature. Among these data, if arranged in a list, there will be found opposite to the element sulphur the number 16, and opposite to lead, 104 *; and since 20 is to 130 in the exact proportion of 16 to 104, it appears that the combination in question affords a satisfactory verification of the law.

(222.) The great importance of physical data of this description, and the advantage of having them well determined, will be obvious, if we consider, that a list of them, when taken in combination with the general law, affords the means of determining at once the exact proportion of the ingredients of all natural compounds, if we only know the place they hold in the system. In chemistry, the number of admitted elements is between fifty and sixty, and new ones are added continually as the science advances. Now, the mo-

* Thomson's First Principles of Chemistry.

ment the number corresponding to any new substance added to the list is determined, we have, in fact, ascertained all the proportions in which it can enter into combination with all the others, so that a careful experiment made with the object of determining this number is, in fact, equivalent to as many different experiments as there are binary, ternary, or yet more complicated combinations capable of existing, into which the new substance may enter, as an ingredient.

(223.) The importance of obtaining exact physical data can scarcely be too much insisted on, for without them the most elaborate theories are little better than mere inapplicable forms of words. It would be of little consequence to be informed, abstractedly, that the sun and planets attract each other, with forces proportional to their masses, and inversely as the squares of their distances: but, as soon as we know the data of our system, as soon as we have an accurate statement (no matter how obtained) of the distances, masses, and actual motions of the several bodies which compose it, we need no more to enable us to predict all the movements of its several parts, and the changes that will happen in it for thousands of years to come; and even to extend our views backwards into time, and recover from the past, phenomena, which no observation has noted, and no history recorded, and which yet (it is possible) may have left indelible traces of their existence in their influence on the state of nature in our own globe, and those of the other planets.

(224.) The proof, too, that our data *are* correctly

assumed, is involved in the general verification of the whole theory, of which, when once assumed, they form a part; and the same comparison with observation which enables us to decide on the truth of the abstract principle, enables us, at the same time, to ascertain whether we have fixed the values of our data in accordance with the actual state of nature. If not, it becomes an important question, whether the assumed values can be corrected, so as to bring the results of theory to agree with facts? Thus it happens, that as theories approach to their perfection, a more and more exact determination of data becomes requisite. Deviations from observed fact, which, in a first or approximative verification, may be disregarded as trifling, become important when a high degree of precision is attained. A difference between the calculated and observed places of a planet, which would have been disregarded by Kepler in his verification of the law of elliptic motion, would now be considered fatal to the theory of gravity, unless it could be shown to arise from an erroneous assumption of some of the numerical data of our system.

(225.) The observations most appropriate for the ready and exact determination of physical data are, therefore, those which it is most necessary to have performed with exactness and perseverance. Hence it is, that their performance, in many cases, becomes a national concern, and observatories are erected and maintained, and expeditions despatched to distant regions, at an expense which, to a superficial view, would appear most disproportioned to their objects But it may very reasonably be asked why the direct assistance afforded by governments to the execution of continued series of observations adapted to this especial end should continue to be, as it has hitherto almost exclusively been, confined to astronomy.

(226.) Physical data intended to be employed as elements of calculation in extensive theories, require to be known with a much greater degree of exactness than any single observation possesses, not only on account of their dignity and importance, as affording the means of representing an indefinite multitude of facts; but because, in the variety of combinations that may arise, or in the changes that circumstances may undergo, cases will occur when any trifling error in one of the data may become enormously magnified in the final result to be compared with observation. Thus, in the case of an eclipse of the sun, when the moon enters very obliquely upon the sun's disc, a trifling error in the diameter of either the sun or moon may make a great one in the time when the eclipse shall be announced to commence. It ought to be remarked, that these are of all others, the conjunctures where observations are most available for the determination of data; for, by the same rule that a small change in the data will, in such cases, produce a great one in the thing to be observed; so, *vice versâ*, any moderate amount of error, committed in an observation undertaken for ascertaining its value, can produce but a very trifling one in the *reverse* calculation from which the data come to be determined by observation. This remark extends to every description of physical data in every department of science, and is never to be overlooked when the object in view is the determination of data with the last degree of precision.

(227.) But how, it may be asked, are we to ascertain *by* observation, data more precise than observation itself? How are we to conclude the value of that which we do not see, with greater certainty than that of quantities which we actually see and measure? It is the number of observations which may be brought to bear on the determination of data that enables us to do this. Whatever error we may commit in a single determination, it is highly improbable that we should always err the same way, so that, when we come to take an average of a great number of determinations, (unless there be some constant cause which gives a bias one way or the other,) we cannot fail, at length, to obtain a very near approximation to the truth, and, even allowing a bias, to come much nearer to it than can fairly be expected from any single observation, liable to be influenced by the same bias.

(228.) This useful and valuable property of the average of a great many observations, that it brings us nearer to the truth than any single observation can be relied on as doing, renders it the most constant resource in all physical enquiries where accuracy is desired. And it is surprising what a rapid effect, in equalizing fluctuations and destroying deviations, a moderate multiplication of individual observations has. A better example can hardly be taken than the average height of the quicksilver in the common barometer, which measures the pressure of the air, and whose fluctuations are proverbial. Nevertheless, if we only observe it regularly every day, and, at the end of each month, take an average of the observed heights, we shall find the fluctuations surprisingly diminished in amount; and if we go on for a whole year, or for many years in succession, the annual averages will be found to agree with still greater exactness. This equalizing power of averages, by destroying all such fluctuations as are irregular or accidental, frequently enables us to obtain evidence of fluctuations really regular, periodic in their recurrence, and so much smaller in their amount than the accidental ones, that, but for this mode of proceeding, they never would have become apparent. Thus, if the height of the barometer be observed four times a day, constantly, for a few months, and the averages taken, it will be seen that a regular *daily* fluctuation, of very small amount, takes place, the quicksilver rising and falling twice in the four-and-twenty hours. It is by such observations that we are enabled to ascertain — what no single measure (unless by a fortunate coincidence), could give us any idea, and never any certain knowledge of — the true *sea level* at any part of the coast, or the height at which the water of the ocean would stand, if perfectly undisturbed by winds, waves, or tides: a subject of very great importance, and upon which it would be highly desirable to possess an extensive series of observations, at a great many points on the coasts of the principal continents and islands over the whole globe.

(229.) In all cases where there is a direct and simple relation between the phenomenon observed and a single *datum* on which it depends, every single observation will give a value of this quantity, and the average of all (under certain restrictions) will be its exact value. We say, under certain restrictions; for, if the circumstances under which the observations are made be not alike, they may not all be equally favourable to exactness, and it would be doing injustice to those most advantageous, to class them with the rest. In such cases as these, as well as in cases where the *data* are numerous and complicated together, so as not to admit of single, separate determination (a thing of continual occurrence), we have to enter into very nice, and often not a little intricate, considerations respecting the *probable* accuracy of our results, or the limits of error within which it is *probable* they lie. In so doing we are obliged to have recourse to a refined and curious branch of mathematical enquiry, called the doctrine of probabilities, the object of which (as its name imports) is to reduce our estimation of the probability of any conclusion to calculation, so as to be able to give more than a mere guess at the degree of reliance which ought to be placed in it.

(230.) To give some general idea of the considerations which such computations involve, let us imagine a person firing with a pistol at a wafer on a wall ten yards distant: we might, in a general way, take it for granted, that he would hit the wall, but not the wafer, at the first shot; but if we would form any thing like a probable conjecture of *how near* he would come to it, we must first have an idea of his skill. No better way of judging could be devised than by letting him fire a hundred shots at it, and marking where they all struck. Suppose this done, — suppose the wafer has been hit once or twice, that a certain number of balls have hit the wall within an inch of it, a certain number between one and two inches, and so on, and that one or two have been some feet wide of the mark. Still the question arises, what estimate are we thence to form of his skill? how *near* (or nearer) may we, after this experience, safely, or at least not unfairly, bet that he will come to the mark the next subsequent shot? This the laws of probability enable us on such data to say. Again, suppose, *before* we were allowed to measure the distances, the wafer were to have been taken away, and we were called upon, on the mere evidence of the marks on the wall, to say where it had been placed; it is clear that no reasoning would enable any one to say with certainty; yet there is assuredly one place which we may fix on with greater probability of being right than any other. Now, this is a very similar case to that of an observer — an astronomer for example — who would determine the exact place of a heavenly body. He points to it his telescope, and obtains a series of results disagreeing among themselves, but yet all agreeing within certain limits, and only a comparatively small number of them deviating considerably from the mean of all; and from these he is called upon to say, definitively, what he shall consider to have been the most probable place of his star at the moment. Just so in the calculation of physical *data;* where no two results agree exactly, and where all come within limits, some wide, some close, what have we to guide us when we would make up our minds what to conclude respecting them? It is evident that any system of calculation that can be shown to lead of necessity to the most probable conclusion where certainty is not to be had must be valuable. However, as this doctrine is one of the most difficult and delicate among the applications of mathematics to natural philosophy, this slight mention of it must suffice at present.

(231.) In the foregoing pages we have endeavoured to explain the spirit of the methods to which, since the revival of philosophy, natural science has been indebted for the great and splendid advances it has made. What we have all along most earnestly desired to impress on the student is, that natural philosophy is essentially united in all its departments, through all which one spirit reigns and one method of enquiry applies. In cannot, however, be studied as a whole, without subdivision into parts; and, in the remainder of this discourse, we shall therefore take a summary view of the progress which has been made in the different branches into which it may be most advantageously so subdivided, and endeavour to give a general idea of the nature of each, and of its relations to the rest. In the course of this, we shall have frequent opportunity to point out the influence of those general principles we have above endeavoured to explain, on the progress of discovery. But this we shall only do as cases arise, without entering into any regular analysis of the history of each department with that view. Such an analysis would, indeed, be a most useful and valuable work, but would far exceed our present limits. We are not, however, without a hope that this great desideratum in science will, ere long, be supplied from a quarter every way calculated to do it justice.

PART III.

OF THE SUBDIVISION OF PHYSICS INTO DISTINCT BRANCHES, AND THEIR MUTUAL RELATIONS.

CHAPTER I.

OF THE PHENOMENA OF FORCE, AND OF THE CONSTITUTION OF NATURAL BODIES.

(232.) NATURAL HISTORY may be considered in two very different lights: either, 1st, as a collection of facts and objects presented by nature, from the examination, analysis, and combination of which we acquire whatever knowledge we are capable of attaining both of the order of nature, and of the agents she employs for producing her ends, and from which, therefore, all sciences arise; or, 2dly, as an assemblage of phenomena to be explained; of effects to be deduced from causes; and of materials prepared to our hands, for the application of our principles to useful purposes. Natural history, therefore, considered in the one or the other of these points of view, is either the beginning or the end of physical science. As it offers to us, in a confused and interwoven mass, the elements of all our knowledge, our business is to disentangle, to arrange, and to present them in a separate and distinct state: and to this end we are called upon to resolve the important but complicated problem,—Given the effect, or assemblage of effects, to find the causes. The principles on which this enquiry relies are those which constitute the relation of cause and effect, as it exists with reference to our minds; and their rules and mode of application have been attempted to be sketched out, (though in far less detail than the intrinsic interest of the subject, both in a logical and practical point of view, would demand,) in the foregoing pages. It remains now to bring together, in a summary statement, the results of the general examination of nature, so far as it has been prosecuted to the discovery of natural agents, and the mode in which they act.

(233.) The first great agent which the analysis of natural phenomena offers to our consideration, more frequently and prominently than any other, is force. Its effects are either, 1st, to counteract the exertion of opposing force, and thereby to maintain *equilibrium*; or, 2dly, to produce *motion* in matter.

(234.) Matter, or that, whatever it be, of which all the objects in nature which manifest themselves directly to our senses consist, presents us with two general qualities, which at first sight appear to stand in contradiction to each other—activity and inertness. Its activity is proved by its power of spontaneously setting other matter in motion, and of itself obeying their mutual impulse, and moving under the influence of its own and other force; inertness, in refusing to move unless obliged to do so by a force impressed externally, or mutually exerted between itself and other matter, and by persisting in its state of motion or rest unless disturbed by some external cause. Yet in reality this contradiction is only apparent.

Force being the cause, and motion the effect produced by it on matter, to say that matter is inert, or has *inertia*, as it is termed, is only to say that the cause is expended in producing its effect, and that the same cause cannot (without renewal) produce double or triple its own proper effect. In this point of view, equilibrium may be conceived as a continual production of two opposite effects, each undoing at every instant what the other has done.

(235.) However, if this should appear too metaphysical, at all events this difference of effects gives rise to two great divisions of the science of force, which are commonly known by the names of STATICS and DYNAMICS; the latter term, which is general, and has been used by us before in its general sense, being usually confined to the doctrine of motion, as produced and modified by force. Each of these great divisions again branches out into distinct subdivisions, according as we consider the equilibrium or motion of matter in the three distinct states in which it is presented to us in nature, the solid, liquid, and aëriform state, to which, perhaps, ought to be added the *viscous*, as a state intermediate between that of solidity and fluidity, the consideration of which, though very obscure and difficult, offers a high degree of interest on a variety of accounts.

(236.) The principles both of the statical and dynamical divisons of mechanics have been definitively fixed by Newton, on a basis of sound induction; and as they are perfectly general, and apply to every case, they are competent, as we have already before observed, to the solution of every problem that can occur in the deductive processes, by which phenomena are to be explained, or effects calculated. Hence, they include every question that can arise respecting the motions and rest of the smallest particles of matter, as well as of the largest masses. But the mode of reasoning from these general principles differs materially, whether we consider them as applied to masses of matter of a sensible size, or to those excessively minute, and perhaps indivisible, molecules of which such masses are composed. The investigations which relate to the latter subject are extremely intricate, as they necessarily involve the consideration of the hypotheses we may form respecting the intimate constitution of the several sorts of bodies above enumerated.

(237.) On the other hand, those which respect the equilibrium and motions of sensible masses of matter are happily capable of being so managed as to render unnecessary the adoption of any particular hypothesis of structure. Thus, in reasoning respecting the application of forces to a solid mass, we suppose its parts indissolubly and unalterably connected; it matters not by what tie, provided this condition be satisfied, that one point of it cannot be moved without setting all the rest in motion, so that the relative situation of the parts one among another be not changed. This is the abstract notion of a solid which the mechanician employs in his reasonings. And their conclusions will apply to natural bodies, of course, only so far as they conform to such a definition. In strictness of speaking, however, there are no bodies which absolutely conform to it. No substance is known whose parts are absolutely incapable of yielding one among another; but the amount by which they do yield is so excessively small as to be demonstrably incapable, in most cases, of having any influence on the results: and in those where it has such influence, an especial investigation of its amount can always be made. This gives rise to two subdivisions of the application of mechanical reasonings to solid masses. Those which refer to the action of forces on flexible or elastic, and on inflexible or rigid, bodies, comprehending under the latter all such whose resistance to flexure or fracture is so very great as to permit our adoption of the language and ideas of the extreme case without fear of material error.

(238.) In like manner, when we reason respecting the action of forces on a fluid mass, all we have occasion to assume is, that its parts are freely moveable one among the other. If, besides this, we choose to regard a fluid as incompressible, and deduce conclusions on this supposition, they will hold good only so far as there may be found such fluids in nature. Now, in strictness, there are none such; but, practically speaking, in the greater number of cases their resistance to compression is so very great that the result of the reasoning so carried on is not sensibly vitiated; and, in the remaining cases, the same general principles enable us to enter on a special enquiry directed to this point: and hence the division of fluids, in mechanical language, into compressible and incompressible, the latter being only the extreme or limiting case of the former.

(239.) As we propose here, however, only to consider what is the actual constitution of nature, we shall regard all bodies, as they really are, more or less flexible and yielding. We know for certain, that the space which any material body appears to occupy is not entirely filled by it; because there is none which by the application of a sufficient force may not be *compressed* or forced into a smaller space, and which, either wholly, as in air or liquids, or in part, as in the greater number of solids, will not recover its former dimensions when the force is taken off. In the case of air, this condensation may be urged to almost any extent; and not only does a mass of air so condensed completely recover its original bulk, when the applied pressure is removed, but if that ordinary pressure under which it exists at the earth's surface (and which arises from the weight of the atmosphere) be also removed by an air-pump, it will still further dilate itself without limit so far as we have yet been able to try it. Hence we are led to the conclusion that the particles of air are mutually elastic, and have a *tendency to recede from one another*, which can only be counteracted by *force*, and therefore is itself a force of the repulsive kind. Nevertheless, as air is heavy, and as gravitation is a universal property of matter, there is no doubt that this repulsive tendency must have a limit, and that there is a distance to which, if the particles of the air could be removed from each other, their mutual repulsion would cease, and an attraction take its place. This limit is probably attained at some very great height above the earth's surface, beyond which, of course, its atmosphere cannot extend.

(240.) What, however, we can only conclude by this or similar reasoning respecting air, we see distinctly in liquids. They are all, though in a small degree, compressible, and recover their former dimensions completely when the pressure is removed; but they cannot be dilated (by mechanical means), and have no tendency, while they remain liquids, to enlarge themselves beyond a certain limit, and therefore they assume a determinate *surface* while at rest, and their parts actually resist further separation with a considerable force, thus giving rise to the phenomenon of the *cohesion of liquids.*

(241.) Both in air and in liquids, however, the most perfect freedom of motion of the parts among each other subsists, which could hardly be the case if they were not separate and independent of each other. And from this, combined with the foregoing considerations, it has been concluded that they do not actually touch, but are kept asunder at determinate distances from each other, by the constant action of the two forces of attraction and repulsion, which are supposed to balance and counteract each other at the ordinary distances of the particles, but to prevail, the one, or the other, according as they are forcibly urged together or pulled asunder.

(242.) In solids, however, the case is very different. The mutual free motion of their parts *inter se* is powerfully impeded, and in some almost destroyed. In some, a slow and gradual change of figure may be produced to a great extent, by pressure or blows, as for instance in the metals, clay, butter, &c.; in others, fracture is the consequence of any attempt to change the figure by violence beyond a certain very small limit. In solids, then, it is evident, that the consideration of their intimate structure has a very great influence in modifying the general results of the action of such attractive and repulsive forces as may be assumed to account for the phenomena they present; yet the general facts that their parts *cohere* with a certain energy, and that they resist displacement or intrusion on the part of other bodies, are sufficient to demonstrate at least the existence of such forces, whatever obscurity may subsist as to their mode of action.

(243.) This division of bodies into airs, liquids, and solids, gives rise, then, to three distinct branches of mechanical science, in each of which the general principles of equilibrium and motion have their peculiar mode of application; viz. pneumatics, hydrostatics, and what might, without impropriety, be termed stereostatics.

Pneumatics.

(244.) Pneumatics relates to the equilibrium or movements of aerial fluids under all circumstances of pressure, density, and elasticity. The weight of the air, and its pressure on all the bodies on the earth's surface, were quite unknown to the ancients, and only first perceived by Galileo, on the occasion of a sucking-pump refusing to draw water above a certain height. Before his time it had always been supposed that water rose by suction in a pipe, in consequence of a certain natural *abhorrence of a vacuum* or empty space, which obliged the water to enter by way of supplying the place of the air sucked out. But if any such abhorrence existed, and had the force of an *acting cause*, which could urge water a single foot into a pipe, there is no reason why the same principle should not carry it up two, three, or any number of feet; none why it should suddenly stop short at a certain height, and refuse to rise higher, however violent the suction might be, nay, even fall back, if purposely forced up too high.

(245.) Galileo, however, at first contented himself with the conclusion, that the natural abhorrence of a vacuum was not strong enough to sustain the water more than about thirty-two feet above its level; and, although the true cause of the phenomenon at length occurred to him, in the pressure of the air on the general surface, it was not satisfactorily demonstrated till his pupil, Torricelli, conceived the happy idea of instituting an experiment on a small scale by the use of a much heavier liquid, mercury, instead of water, and, in place of sucking out the air from above, employing the much more effectual method of filling a long glass tube with mercury, and inverting it into a basin of the same metal. It was then at once seen, as by a *glaring instance*, that the maintenance of the mercury in the tube (which is nothing else than the common barometer) was the effect of a perfectly definite external cause, while its fluctuations from day to day, with the varying state of the atmosphere, strongly corroborated the notion of its being due to the pressure of the external air on the surface of the mercury in the reservoir.

(246.) The discovery of Torricelli was, however, at first much misconceived, and even disputed, till the question was finally decided by appeal to a *crucial instance*, one of the first, if not the very first on record in physics, and for which we are indebted to the celebrated Pascal. His acuteness perceived that if the weight of the incumbent air be the direct cause of the elevation of the mercury, it must be measured by the amount of that elevation, and therefore that, by carrying a barometer up a high mountain, and so ascending into the atmosphere *above* a large portion of the incumbent air, the pressure, as well as the length of the column sustained by it, must be diminished; while, on the other hand, if the phenomenon were due to the cause originally assigned, no difference could be expected to take place, whether the observation were made on a mountain or on the plain. Perhaps the decisive effect of the experiment which he caused to be instituted for the purpose, on the Puy de Dôme, a high mountain in Auvergne, while it convinced every one of the truth of Torricelli's views, tended more powerfully than any thing which had previously been done in science to confirm, in the minds of men, that disposition to experimental verification which had scarcely yet taken full and secure root.

(247.) Immediately on this discovery followed that of the air-pump, by Otto von Guericke of Magdeburgh, whose aim seems to have been to decide the question, whether a vacuum could or could not exist, by endeavouring to make one. The imperfection of his mechanism enabled him only to diminish the aërial contents of his receivers, not entirely to empty them; but the curious effects produced by even a partial exhaustion of air speedily excited attention, and induced our illustrious countryman, Robert Boyle, to the prosecution of those experiments which terminated in his hands, and in those of Hauksbee, Hooke, Mariotte, and others, in a satisfactory knowledge of the general law of the equilibrium of the air under the influence of greater or less pressures. These discoveries have since been extended to all the various descriptions of aërial fluids which chemistry has shown to exist, and to maintain their aëriform state under artificial pressure, and even to those which may be produced from liquids reduced to a state of vapour by heat, so long as they retain that state.

(248.) The manner in which the observed law of equilibrium of an elastic fluid, like air, may be considered to originate in the mutual repulsion of its particles, has been investigated by Newton, and the actual statement of the law itself, as announced by Mariotte, "that the density of the air, or the quantity of it contained in the same space, is, *cæteris paribus*, proportional to the pressure it supports," has recently been verified within very extensive limits by direct experiment, by a committee of the Royal Academy of Paris. This law contains the principle of solution of every dynamical question that can occur relative to the equilibrium of elastic fluids, and is therefore to be regarded as one of the highest *axioms* in the science of pneumatics.

Hydrostatics.

(249.) The principles of the equilibrium of liquids, understanding by this word such fluids as do not, though quite at liberty, attempt to dilate themselves beyond a certain point, are at once few and simple. The first steps towards a knowledge of them were made by Archimedes, who established the general fact, that a solid immersed in a liquid loses a portion of its weight equal to that of the liquid it displaces. It seems very astonishing, after this, that it should not have been at once concluded that the weight thus said to be *lost* is only *counteracted* by the upward pressure of the liquid, and that, therefore, a portion of any liquid, surrounded on all sides by a liquid of the same kind, does really exert its weight in keeping its place. Yet the prejudice that "liquids do not gravitate in their natural place" kept its ground, and was only dispelled with the mass of error and absurdity which the introduction of a rational and experimental philosophy by Galileo swept away.

(250.) The hydrostatical law of *the equal pressure of liquids in all directions*, with its train of curious and important consequences, is an immediate conclusion from the perfect mobility of their parts among one another, in consequence of which each of them tends to recede from an excess of pressure on one side, and thus bears upon the rest, and distributes the pressure among its neighbours. In this form it was laid down by Newton, and has proved one of the most useful and fertile principles of physico-mathematical reasoning on the equilibrium of fluid masses, as affording a means of tracing the action of a force applied at any point of a liquid through its whole extent. It applies, too, without any modification, to expansible fluids as well as to liquids; and, in the applications of geometry to this subject, enables us to dispense with any minute and intricate enquiries as to the mode in which individual particles act on each other.

(251.) In a practical point of view, this law is

remarkable for the directness of its application to useful purposes. The immediate and perfect distribution of a pressure applied on any one part, however small, of a fluid surface through the whole mass, enables us to communicate *at one instant* the same pressure to any number of such parts by merely increasing the surface of the fluid, which may be done by enlarging the containing vessel; and if the vessel be so constructed that a large portion of its surface shall be moveable together, the pressures on all the similar parts of this portion will be united into one consentient force, which may thus be increased to any extent we please. The hydraulic press, invented by Bramah, (or rather applied by him after a much more ancient inventor, Stevin,) is constructed on this principle. A small quantity of water is driven by sufficient pressure into a vessel *already full*, and provided with a moveable surface or piston of great size. Under such circumstances something must give way; the great surface of the piston accumulates the pressure on it to such an extent that nothing can resist its violence. Thus, trees are torn up by the roots; piles extracted from the earth; woollen and cotton goods compressed into the most portable dimensions; and even hay, for military service, reduced to such a state of coercion as to be easily packed on board transports.

(252.) Liquids differ from aëriform fluids by their *cohesion*, which may be regarded as a kind of approach to a solid state, and was so regarded by Bacon (193.). Indeed, there can be little doubt that the solid, liquid, and aëriform states of bodies are merely stages in a progress of gradual transition from one extreme to the other; and that, however strongly marked the distinctions between them may appear, they will ultimately turn out to be separated by no sudden or violent line of demarcation, but shade into each other by insensible gradations. The late experiments of Baron Cagnard de la Tour may be regarded as a first step towards the full demonstration of this (199.). But the cohesion of liquids is not, like that of solids, so modified by their structure in other respects as to destroy the mobility of their parts one among another (unless in those cases of nearer approach to the solid state which obtain in viscid or gummy liquids). On the contrary, the two qualities co-exist, and give rise to a number of curious and intricate phenomena.

(253.) One of the most remarkable of these is capillary attraction, or capillarity as it is sometimes called. Every body has remarked the adhesion of water to glass. The elevation of the general surface of the liquid where it is in contact with the containing vessel; the form of a drop suspended at the under side of a solid: these are instances of capillary attraction. If a small glass tube with a bore as fine as a hair be immersed in water, the water will be observed to rise in it to a certain height, and to assume a concave surface at its upper extremity. The attraction of the glass on the water, and the cohesion of the parts of the water to each other, are no doubt the joint causes of this curious effect; but the mode of action is at once obscure and complex; and although the researches of Laplace and Young have thrown great light on it, further investigation seems necessary before we can be said distinctly to understand it.

(254.) As the capillarity and cohesion of the parts of liquids shows them to possess the power of mutual attraction, so their elasticity demonstrates that they also possess that of repulsion when forcibly brought nearer than their natural state. From the extremely small extent to which the compression of liquids can be carried by any force we can employ, compared with that of air, we must conclude that this repulsion is much more violent in the former than in the latter, but counteracted also by a more powerful force of attraction. So much more powerful, indeed, is the resistance of liquids to compression, that they were usually regarded as incompressible; an opinion corroborated by a celebrated experiment made at Florence, in which water was forced through the pores (as it was said) of a golden ball. More recent experiments by Canton, and since by Perkins, Oersted, and others, have demonstrated however the contrary, and assigned the amount of compression.

(255.) The consideration of the motions of fluids, whether liquid or expansible, is infinitely more complicated than that of their equilibrium. When their motions are slow, it is reasonable to suppose that the law of the equable distribution of pressure obtains; but in very rapid displacements of their parts one among the other, it is not easy to see how such an equable distribution can be accomplished, and some phenomena exist which seem to indicate a contrary conclusion.

(256.) Independent of this, there are difficulties

of an almost insuperable nature to the regular deductive application of the general principles of mechanics to this subject, which arise from the excessive intricacy of the pure mathematical enquiries to which its investigation leads. It was Newton who set the example of a first attempt to draw any conclusions respecting the motion of fluid masses by direct reasoning from dynamical principles, and thus laid the foundation of HYDRODYNAMICS; but it was not till the time of D'Alembert that the method of reducing any question respecting the motions of fluids under the action of forces to strict mathematical investigation could be said to be completely understood. But the cases even now in which this mode of treating such questions can be applied with full satisfaction are few in comparison of those in which the experimental method of enquiry as already observed (189.) is preferable. Such, for example, is that of the resistance of fluids to bodies moving through them; a knowledge of which is of great importance in naval architecture and in gunnery, where the resistance of the air acts to an enormous extent. Such, too, among the practical subjects which depend mainly on this branch of science, are the use of sails in navigation; the construction of wind-mills, and water-wheels; the transmission of water through pipes and channels; the construction of docks and harbours, &c.

Nature of Solids in general.

(257.) The intimate constitution of solids is, in all probability, very complicated, and we cannot be said to know much of it. By some recent delicate experiments on the dimensions of wires violently strained, it has been shown that they are to a certain small extent capable of being dilated by tension, as they are also of being compressed by pressure, but within limits even narrower than those of liquids. Usually, when strained too far, they break, and refuse to re-unite; or, if compressed too forcibly, take a permanent contraction of dimension. Thus, wood may be indented by a blow, and metals rendered denser and heavier by hammering or rolling. There is a certain degree of confusion prevalent in ordinary language about the hardness, elasticity, and other similar qualities, of solids, which it may be well to remove. Hardness is that disposition of a solid which renders it difficult to displace its parts among themselves. Thus, steel is harder than iron; and diamond almost infinitely harder than any other substance in nature: but the compressibility of steel, or the extent to which it will yield to a given pressure and recover itself, is not much less than that of soft iron, and that of ice is very nearly the same with that of water.

(258.) Again, we call Indian rubber a very elastic body, and so it is; but in a different sense from steel. Its parts admit of great mutual displacement without permanent dislocation; however distorted, it recovers its figure readily, but with a small force. Yet, if Indian rubber were to be enclosed in a space that it just filled, so as not to permit its parts to yield laterally, doubtless it would resist actual compression with great violence. Here, then, we have an instance of two kinds of elasticity in one substance; a feebler effort of recovery from distorted figure, and a more violent one from a state of altered dimension. Both, however, originate in the same causes, and are referable to the same principles; the former being in fact only a modified case of the latter, as the effort of a steel spring, when bent, to recover its former shape, is referable to the same forces which give to steel its hardness and strength to resist actual compression and fracture.

(259.) The toughness of a solid, or that quality by which it will endure heavy blows without breaking, is again distinct from hardness though often confounded with it. It consists in a certain yielding of parts with a powerful general cohesion, and is compatible with various degrees of elasticity. Malleability is again another quality of solids, especially metals, quite distinct from toughness, and depends on their capability of being deprived of their figure without an effort to recover it and without fracture.

(260.) Tenacity, again, is a property of solids more directly depending on the cohesion of their parts than toughness. It consists in their power of resisting separation by a strain steadily applied, while the quality of toughness is materially influenced by their disposition to communicate through their substance the jarring effect of a blow. Accordingly, the tenacity of a solid is a direct measure of the cohesive attraction of its parts, and is the best proof of the existence of such a power.

Crystallography.

(261.) It cannot be supposed that these and many other tangible qualities, as they may be called, should subsist in solids without a corresponding mechanism in their internal structure. That they have such a mechanism, and that a very curious and intricate one, the phenomena of crystallography sufficiently show. This interesting and beautiful department of natural science is of comparatively very modern date. That many natural substances affected certain forms must have been known from the earliest times. Pliny appears to have been acquainted with this fact, at least in some instances, as he describes the forms of quartz and diamond. But till the time of Linnæus no material attention seems to have been bestowed on the subject. He, however, observed, and described with care, the crystalline forms of a variety of substances, and even regarded them as so definite a character of the solids which assumed them, that he supposed every particular form to be generated by a particular salt. Romé de l'Isle pursued the study of the crystalline forms of bodies yet farther. He first ascertained the important fact of the constancy of the angles at which their faces meet; and observing further that many of them appear in several different shapes, first conceived the idea that these shapes might be reducible to one, appropriated in a peculiar manner to each *substance*, and modified by strict geometrical laws. Bergmann, reasoning on a fact imparted to him by his pupil Gahn, made a yet greater step, and showed how at least one species of crystal might be built up of thin laminæ ranged in a certain order, and following certain rules of superposition. He failed, however, in deducing just and general conclusions from this remark, which, correctly viewed, is the foundation of the most important law of crystallography, that which connects the primitive form with other forms capable of being exhibited by the same substance, by a certain fixed relation. An idea may be formed of what is meant by this sort of connection of one form with another, by considering a pointed pyramid built up of cubic stones, disposed in layers, each of which separately is a square plate of the thickness of a single stone. These layers, laid horizontally one on the other, and decreasing regularly in size from the bottom to the top, produce a pyramidal form with a rough or channeled surface; and if the layers are so extremely thin that the channels cease to be visible to the eye, the pyramid will seem smooth and perfect.

(262.) Very shortly after this, and without knowledge of what had been done by Gahn and Bergmann, the Abbé Haüy, instructed by the accidental fracture of a fine groupe of crystals, made the remark noticed already (in 67.), and reasoning on it with more caution and success, and pursuing it into all its detail, developed the general laws which regulate the superposition of the layers of particles of which he supposes all crystals to be built up, and which enable us, from knowing their primitive forms, to discover, previous to trial, what other forms they are capable of assuming; and which, according to this idea, are called derivative or secondary forms. Mohs and others have since imagined processes and systems by which the derivation of forms from each other is facilitated, and have corrected some errors of over-hasty generalization into which their predecessors had fallen, as well as advanced, by an extraordinary diligence of research, our knowledge of the forms which the various substances which occur in nature and art actually do assume.

(263.) In what manner a variety in point of external form may originate in a variety of figures in the ultimate particles of which a solid is composed, may very readily be imagined by considering what would happen if the bricks of which an edifice is constructed had all a certain *leaning* or bias in one direction out of the perpendicular. Suppose every brick, for instance, when laid flat on its face, with its longer edges north and south, had its eastern and western faces upright, but its northern and southern ones leaning southwards at a certain inclination the same for each brick; a house built of such bricks would lean the same way, if the bricks fitted well together. If, *besides this*, the eastern and western faces of the bricks, instead of being truly upright, had an inclination eastward, the house would have a similar one, and all its four corners, instead of being upright, would lean to the south-east. Suppose, instead of a house, a pyramid were built of such oblique bricks, with the sides of its base directed to the four points of the compass; then its point, instead of being situated vertically over the centre of its base, would stand perpendicularly over some point to the south-east of that centre, and the pyramid itself would have its sides facing the south and the east, more highly inclined to the horizon than those towards the north and west.

(264.) Whatever conception we may form of the manner in which the particles of a crystal cohere and form masses, it is next to impossible to divest ourselves of the idea of a determinate figure common to them all. Any other supposition, indeed, would be incompatible with that exact similarity in all other respects which the phenomena of chemistry may be considered as having demonstrated. However, it must be borne in mind that this idea, plausible as it may appear, is yet in some degree hypothetical, and that the laws of crystallography, as determined from inductive observation, are quite independent of any supposition of the kind, or even of the existence of such things as ultimate particles or atoms at all.

(265.) Still, that peculiar internal constitution of solid bodies, whatever it be, which is indicated by the assumption of determinate figures, by their splitting easier in some directions than in others, and by their presenting glittering plane surfaces when broken into fragments, cannot but have an important influence on all their relations to external agents, as well as to their internal movements and the mutual actions of their parts on one another. Accordingly, the division of bodies into crystallized and uncrystallized, or imperfectly crystallized, is one of the most universal importance; and almost all the phenomena produced by those more intimate natural causes which act within small limits, and as it were on the immediate mechanism of solid sub. stances, are remarkably modified by their crystalline structure. Thus, in transparent solids, the course taken by the rays of light, in traversing them, as well as the properties impressed upon them in so doing, are intimately connected with this structure The recent experiments of M. Savart, too, have proved that this is also the case with their power of resistance to external force, on which depends their elasticity. Crystallized substances, according to the results of these experiments, resist compression with different degrees of elastic force, according to the direction in which it is attempted to compress them; and all the phenomena dependent on their elasticity are affected by this cause, especially those which relate to their vibratory movements and their conveyance of sound.

(266.) There can be little doubt that modifications, similarly depending on the internal structure of crystals, will be traced through every department of physics. In that interesting one which relates to the action of heat in expanding the dimensions of substances, a beginning has already been made by Professor Mitscherlich. It had long been known that all substances are dilated by heat, and no exception to this law has been found, so long as we regard the *bulk* of the heated body. Thus, an iron rod when hot is both longer and thicker than when cold; and the difference of dimension, though but trifling in itself, is yet capable of being made sensible, and is of considerable consequence in engineering. Thus, too, the quicksilver in a common thermometer occupies a larger space when hot than when cold; and being confined by the glass ball, (which also expands, but *not so much in proportion,*) it is forced to rise in the tube. These and similar facts had long been known; and accurate measures of the total amount of dilatation of a variety of different bodies, under similar accessions of heat, had been obtained and registered in tables. But no one had suspected the important fact, that this expansion in crystallized bodies takes place under totally different circumstances from what obtains in uncrystallized ones. M. Mitscherlich has lately shown that such substances expand differently in different directions, and has even produced a case in which expansion in one direction is actually accompanied with contraction in another. This step, the most important beyond a doubt which has yet been made in pyrometry, can however only be regarded as the first in a series of researches which will occupy the next generation, and which promises to afford an abundant harvest of new facts, as well as the elucidation of some of the most obscure and interesting points in the doctrine of heat.

(267.) From what has been said, it is clear that if we look upon solid bodies as collections of particles or atoms, held together and kept in their places by the perpetual action of attractive and repulsive forces, we cannot suppose these forces, at least in crystallized substances, to act alike in all directions. Hence arises the conception of *polarity*, of which we see an instance, on a great scale, in the magnetic needle, but which, under modified forms, there is nothing to prevent us from conceiving to act among the ultimate atoms of solid or even fluid bodies, and to produce all the phenomena which they exhibit in their crystallized state, either when acting on each other, or on light, heat, &c. It is not difficult, if we give the reins to imagination, to conceive how attractive and repulsive atoms, bound together by some unknown tie, may form little machines or compound particles, which shall have many of the properties which we refer to polarity; and accordingly many ingenious suppositions have been made to that effect: but in the actual state of science it is certainly safest to wave these hypotheses, without however absolutely rejecting them, and regard the *polarity of matter* as one of the ultimate phenomena to which the analysis of nature leads us, and of which it is our business fully to investigate the laws, before we endeavour to ascertain its causes, or trace the mechanism by which it is produced.

(268.) The mutual attractions and repulsions of the particles of matter, then, and their polarity, whether regarded as an original or a derivative property, are the forces which, acting with great energy, and within very confined limits, we must look to as the principles on which the intimate constitution of all bodies and many of their mutual actions depend. These are what are understood by the general term of *molecular forces*. Molecular attraction has been attempted to be confounded by some with the general attraction of gravity, which all matter exerts on all other matter; but this idea is refuted by the plainest facts.

CHAP. II.

OF THE COMMUNICATION OF MOTION THROUGH BODIES. — OF SOUND AND LIGHT.

(269.) The propagation of motion through all substances, whether of a single impulse, as a blow or thrust, or of one frequently and regularly repeated, such as a jarring or vibratory movement, depends wholly on these molecular forces; and it is on such propagation that sound and very probably light depend. To conceive the manner in which a motion may be conveyed from one part of a substance to another, whether solid or fluid, we may attend to what takes place when a wave is made to run along a stretched string, or the surface of still water. Every part of the string, or water, is in succession moved from its place, and agitated with a motion similar to that of the original impulse, leaving its place and returning to it, and when one part ceases to move the next receives as it were the impression, and forwards it onward. This may seem a slow and circuitous process in description; but when sound, for example, is conveyed through the air, we are to consider, 1st, that the air, the substance actually in motion, is extremely light and acted upon by a very powerful elasticity, so that the force which propagates the motion, or by which the particles adjacent act on, and urge forward, each other, is very great, compared with the quantity of materials set in motion by it:

and the same is true, even in a greater degree, in liquids and solids; for in these the elastic forces are even greater, in proportion to the weight, than in air.

(270.) A general notion of the mode in which sounds are conveyed through the air was not altogether deficient among the ancients; but it is to Newton that we owe the first attempt to analyze the process, and show correctly what takes place in the communication of motion from particle to particle. Reasoning on the properties of the air as an elastic body, he showed the effect of an impulse on any portion of it to consist in a condensation of the air immediately adjacent in the direction of the impulse, which then, re-acting by its spring, drives back the portion which had advanced to its original place, and at the same time urges forward the portion before it, in the direction of the impulse, so that every particle alternately advances and retreats. But, in pursuing this idea into its details, Newton fell into some errors which were pointed out by Cramer, though their origin was not traced, nor the reasoning corrected, till the subject was resumed by Lagrange and Euler; nor is this any impeachment of the penetration of our immortal countryman. The mathematical theory of the propagation of sound, and of vibratory and undulatory motions in general, is one of the utmost intricacy; and, in spite of every exertion on the part of the most expert geometers, continues to this day to give continual occasion for fresh researches; while phenomena are constantly presenting themselves, which show how far we are from being able to deduce all the particulars, even of cases comparatively simple, by any direct reasoning from first principles.

(271.) Whenever an impulse of any kind is conveyed by the air, to our ears, it produces the impression of sound; but when such an impulse is regularly and uniformly repeated in extremely rapid succession, it gives us that of a musical note, the pitch of the note depending on the rapidity of the succession (see art. 153.). The sense of harmony, too, depends on the periodical recurrence of coincident impulses on the ear, and affords, perhaps, the only instance of a sensation for whose pleasing impression a distinct and intelligible reason can be assigned.

(272.) Acoustics, then, or the science of sound, is a very considerable branch of physics, and one which has been cultivated from the earliest ages. Even Pythagoras and Aristotle were not ignorant of the general mode of its transmission through the air, and of the nature of harmony; but as a branch of science, independent of its delightful application in the art of music, it could be hardly said to exist, till its nature and laws became a matter of experimental enquiry to Bacon and Galileo, Mersenne and Wallis; and of mathematical investigation to Newton, and his illustrious successors, Lagrange and Euler. From that time its progress, as a branch both of mathematical and experimental science, has been constant and accelerated. A curious and beautiful method of observation, due to Chladni, consists in the happy device of strewing sand over the surfaces of bodies in a state of sonorous vibration, and marking the figures it assumes. This has made their motions susceptible of ocular examination, and has been lately much improved on, and varied in its application, by M. Savart, to whom we also owe a succession of instructive researches on every point connected with the subject of sound, which may rank among the finest specimens of modern experimental enquiry. But the subject is far from being exhausted; and, indeed, there are few branches of physics which promise at once so much amusing interest, and such important consequences, in its bearings on other subjects, and especially, through the medium of strong analogies, on that of light.

Light and Vision.

(273.) The nature of light has always been involved in considerable doubt and mystery. The ancients could scarcely be said to have any opinion on the subject, unless, indeed, it could be considered such to affirm that distant bodies could not be put into communication without an intermedium; and that, therefore, there must be *something* between the eye and the thing seen. What that something is, however, they could only form crude and vague conjectures. One supposed that the eyes themselves emit rays or emanations of some unknown kind, by which distant objects are as it were felt; a singularly unfortunate idea, since it gives no reason why objects should not be equally well seen in the dark — no account, in short, of the part performed by *light* in vision. Others imagined that all visible objects are constantly throwing out from them, in all directions, some sort of resemblances or spectral forms of themselves, which, when received by the eyes, produce an impression of the objects. Vague and clumsy as this hypothesis obviously is, it assigns to the object a power, and to light a diffusive propagation in all directions, which are, the one and the other, independent of our eyes, and therefore goes to separate the phenomena of *light* from those of *vision.*

(274.) The hypothesis of Newton is a refinement and improvement on this idea. Instead of spectra or resemblances, he supposes luminous objects actually to dart out from them in all directions, particles of inconceivable minuteness (as indeed they must be, having such an enormous velocity (see 17.), not to dash in pieces every thing they strike upon). These particles he supposes to be acted upon by attractive and repulsive forces, residing in all material bodies, the latter extending to some very small distance beyond their surfaces; and by the action of these forces to be turned aside from their natural straight-lined course, without ever coming in actual contact with the particles themselves of the bodies on which they fall, but either being turned back and *reflected* by the repulsive forces before they reach them, or penetrating between their intervals, as a bird may be supposed to fly through the branches of a forest, and undergoing all their actions, to take at quitting them a direction finally determined by the position of the surface at which they emerge with respect to their course.

(275.) This hypothesis, which was discussed and reasoned upon by Newton in a manner worthy of himself, affords, by the application of the same dynamical laws which he had applied with so much success to the explanation of the planetary motions, not merely a plausible, but a perfectly reasonable and fair explanation of all the *usual* phenomena of light known in his time. His own beautiful discoveries, too, of the different refrangibilities of the differently coloured rays, were perfectly well represented in this theory, by simply admitting a difference of velocity in the particles, which produce in the eye the sensations of different colours. And had the properties of light remained confined to these, there would have been no occasion to have resorted to any other mode of conceiving it.

(276.) A very different hypothesis had, however, been suggested about the same period by Huyghens, who supposed light to be produced in the same manner with sound, by the communication of a vibratory motion from the luminous body to a highly elastic fluid, which he imagined as filling all space, and as being less condensed within the limits of space occupied by matter, and that to a greater or less extent, according to the nature of the occupying substance. Thus, in place of any thing actually thrown off, he substituted waves, or vibrations, propagated in all directions from luminous bodies, through this medium, or ether, as he called it. Huyghens, being himself a consummate mathematician, was enabled to trace many of the consequences of this hypothesis, and to show that the ordinary laws of reflection and refraction were represented or accounted for by it, as well as by Newton's. But the hypothesis of Huyghens has not been fully successful in accounting for what may be considered the chief of all optical facts, the production of colours in the ordinary refraction of light by a prism, of which the theory of Newton gives a complete and elegant explanation; and the discovery of which by him marks one of the greatest epochs in the annals of experimental science. This, which has been often urged in objection to it, remains still, if not quite unanswered, at least only imperfectly removed.

(277.) Other phenomena, however, were not wanting to afford a further trial of the *explanatory powers* of either hypothesis. The diffraction or inflection of light, discovered by Grimaldi, a Jesuit of Bologna, seemed to indicate that the rays of light were turned aside from their straight course by merely passing near bodies of every description. These phenomena, which are very curious and beautiful, were minutely examined by Newton, and referred by him to the action of repulsive forces extending to a sensible distance from the surfaces of bodies; and his explanation, so far as the facts known to him are concerned, appears as satisfactory as could reasonably be then expected; and much more so than any thing which could at that time be produced on the side of the hypothesis of Huyghens, which, in fact, seemed incapable of giving any account whatever of them.

(278.) Another class of delicate and splendid optical phenomena, which had begun to attract attention somewhat previous to Newton's time, seemed to leave both hypotheses equally at a loss. These were the colours exhibited by very thin films, either of a liquid (such as a soap-bubble), or of air, as when two glasses are laid together with only air between them. These colours were examined by Newton with a minuteness and care altogether unexampled in experimental philosophy at that time, and with which few researches undertaken since will bear to stand in competition. Their result was a theory of a very singular nature, which he grounded on an hypothesis of what he termed *fits of easy transmission and reflection;* and which supposed each ray of light to pass in its progress periodically through a succession of states such as would alternately dispose it to penetrate or be reflected back from the surface of a body on which it might fall. The simplest way in which the reader may conceive this hypothesis, is to regard every particle of light as a sort of little magnet revolving rapidly about its own centre while it advances in its course, and thus alternately presenting its attractive and repulsive pole, so that when it arrives at the surface of a body with its repulsive pole foremost, it is repelled and reflected; and when the contrary attracted, so as to enter the surface. Newton, however, very cautiously avoided announcing his theory in this or any similar form, confining himselt entirely to general language. In consequence, it has been confidently asserted by all his followers, that the doctrine of fits of easy reflection and transmission, as laid down by him, is substantially nothing more than a statement of facts. Were it so, it is clear that any other theory which should offer a just account of the same phenomena must ultimately involve and coincide with that of Newton. But this, as we shall presently see, is not the case; and this instance ought to serve to make us extremely cautious how we employ, in stating physical laws derived from experiment, language which involves any thing in the slightest degree theoretical, if we would present the laws themselves in a form which no future research shall modify or subvert.

(279.) A third class of optical phenomena, which were likewise discovered while Newton was yet engaged in his optical researches, was that exhibited by doubly refracting crystals. In what the phenomenon of double refraction consists, we have already had occasion to explain. The fact itself was first noticed by Erasmus Bartolin in the crystal called Iceland spar; and was studied with attention by Huyghens, who ascertained its laws, and referred it with remarkable ingenuity and success to his theory of light, by the additional hypothesis of such a constitution of his ethereal medium within the crystal as should enable it to convey an impulse faster in one direction than another: as if, for example's sake, we should suppose a sound conveyed through the air with different degrees of rapidity in a vertical and horizontal direction.

(280.) Some remarkable facts accompanying the double refraction produced by Iceland spar, which Bartolin, Huyghens, and Newton, had observed, led the latter to conceive the singular idea that a ray of light after its emergence from such a crystal acquires *sides*, that is to say, distinct relations to surrounding space, which it carries with it through its whole subsequent course, and which give rise to all those curious and complicated phenomena which are now known under the name of the *polarization of light*. These results, however, appeared so extraordinary, and offered so little handle for further enquiry, that their examination dropped, as if by common consent; Newton himself resting content with urging strongly the apparent incompatibility of these properties with the Huyghenian doctrine, but without making any attempt to explain them by his own.

(281.) From the period of Newton's optical discoveries to the commencement of the present century, no great accession to our knowledge of the nature of light was made, if we except one, which, from its invaluable practical application, must ever hold a prominent place in the annals both of art and science: we mean, the discovery of the principle of the achromatic telescope, which originated in a discussion between the celebrated geometer Euler, Klingenstierna, an eminent Swedish philosopher, and our own countryman, the admirable optician Dollond, on the occasion of certain abstract theoretical investigations of the former, which led him to speculate on its *possibility*, and which ultimately terminated in its complete and happy *execution* by the latter; a memorable case in science, though not a singular one, where the speculative geometer in his chamber, apart from the world, and existing among abstractions, has originated views of the noblest practical application.*

(282.) The explanation which our knowledge of optical laws affords of the mechanism of the eye, and the process by which vision is performed, is as com-

* There seems no doubt, however, that an achromatic telescope had been constructed by a private amateur, a Mr. Hall, some time before either Euler or Dollond ever thought of it.

plete and satisfactory as that of hearing by the propagation of motion through the air. The camera obscura, invented by Baptista Porta in 1560, gave the first idea how the actual images of external objects might be conveyed into the eye, but it was not till after a considerable interval that Kepler, the immortal discoverer of those great laws which regulate the periods and motions of the planets, pointed out distinctly the offices performed by the several parts of the eye in the act of vision. From this to the invention of the telescope and microscope there would seem but a small step, but it is to accident rather than design that it is due; and its reinvention by Galileo, on a mere description of its effects, may serve, among a thousand similar instances, to show that inestimable practical applications lie open to us, if we can only once bring ourselves to conceive their possibility, a lesson which the invention of the achromatic telescope itself, as we have above related it, not less strongly exemplifies.

(283.) The little instrument with which Galileo's splendid discoveries were made was hardly superior in power to an ordinary finder of the present day; but it was rapidly improved on, and in the hands of Huyghens attained to gigantic dimensions and very great power. It was to obviate the necessity of the enormous length required for these telescopes, and yet secure the same power, that Gregory and Newton devised the reflecting telescope, which has since become a much more powerful instrument than its original inventors probably ever contemplated.

(284.) The telescope, as it exists at present, with

the improvements in its structure and execution which modern artists have effected, must assuredly be ranked among the highest and most refined productions of human art; that in which man has been able to approximate more closely to the workmanship of nature, and which has conferred upon him, if not another sense, at least an exaltation of one already possessed by him that merits almost to be regarded as a new one. Nor does it appear yet to have reached its ultimate perfection, to which indeed it is difficult to assign any bounds, when we take into consideration the wonderful progress which workmanship of every kind is making, and the delicacy, far superior to that of former times, with which materials may now be wrought, as well as the ingenious inventions and combinations which every year is bringing forth for accomplishing the same ends by means hitherto unattempted.*

(285.) After a long torpor, the knowledge of the properties of light began to make fresh progress about the end of the last century, advancing with an accelerated rapidity, which has continued unabated to the present time. The example was set by our late admirable and lamented countryman,

* We allude to the recently invented achromatic combinations of Messrs. Barlow and Rogers, and the dense glasses of which Mr. Faraday has recently explained the manufacture in a memoir full of the most beautiful examples of delicate and successful chemical manipulation, and which promise to give rise to a new era in optical practice, by which the next generation at least may benefit. See Phil. Trans. 1830.

Dr. Wollaston, who re-examined and verified the laws of double refraction in Iceland spar announced by Huyghens. Attention being thus drawn to the subject, the geometry of Laplace soon found a means of explaining at least one portion of the mystery of this singular phenomenon, by the Newtonian theory of light, applied under certain supposed conditions; and the reasoning which led him to the result (at that time quite unexpected), may justly be regarded as one of his happiest efforts. The prosecution of the subject, which had now acquired a high degree of interest, was encouraged by the offer of a prize on the part of the French Academy of Sciences; and it was in a memoir which received this honourable reward on that occasion, in 1810, that Malus, a retired officer of engineers in the French army, announced the great discovery of the *polarization of light* by ordinary reflection at the surface of a transparent body.

(286.) Malus found that when a beam of light is reflected from the surface of such a body at a certain angle, it acquires precisely the same singular property which is impressed upon it in the act of double refraction, and which Newton had before expressed by saying that it possessed *sides*. This was the first circumstance which pointed out a connection between that hitherto mysterious phenomenon and any of the ordinary modifications of light; and it proved ultimately the means of bringing the whole within the limits, if not of a complete explanation, at least of a highly plausible theoretical representation. So true is, in science, the remark of Bacon, that no natural phenomenon can be adequately studied *in itself alone*, but, to be understood, must be considered *as it stands connected with all nature*.

(287.) The new class of phenomena thus disclosed were immediately studied with diligence and success, both abroad by Malus and Arago, and at home by our countryman Dr. Brewster, and their laws investigated with a care proportioned to their importance; when another and apparently still more extraordinary class of phenomena presented itself in the production of the most vivid and beautiful colours (every way resembling those observed by Newton in thin films of air or liquids, only infinitely more developed and striking,) in certain transparent crystallized substances, when divided into flat plates in particular directions, and exposed in a beam of polarized light. The attentive examination of these colours by Wollaston, Biot, and Arago, but more especially by Brewster, speedily led to the disclosure of a series of optical phenomena so various, so brilliant, and evidently so closely connected with the most important points relating to the intimate structure of crystallized bodies, as to excite the highest interest, — that sort of interest which is raised when we feel we are on the eve of some extraordinary discovery, and expect every moment that some leading fact will turn up, which will throw light on all that appears obscure, and reduce into order all that seems anomalous.

(288.) This expectation was not disappointed. So long before the time we are speaking of as the first year of the present century, our illustrious countryman, the late Dr. Thomas Young, had established a principle in optics, which, regarded as a physical law, has hardly its equal for beauty, simplicity, and extent of application, in the whole circle of science. Considering the manner in which the vibrations of two musical sounds arriving at once at the ear affect the sense with an impression of sound or silence according as they conspire or oppose each other's effects, he was led to the idea that the same ought to hold good with light as with sound, if the theory which makes light analogous to sound be the true one; and that, therefore, two rays of light, setting off from the same origin, at the same instant, and arriving at the same place by different routes, ought to strengthen or wholly or partially destroy each other's effects according to the difference in length of the routes described by them. That two lights should in any circumstances combine to produce darkness may be considered strange, but is *literally true;* and it had even been noticed long ago as a singular and unaccountable fact by Grimaldi, in his experiments on the inflection of light. The experimental means by which Dr. Young confirmed this principle, which is known in optics by the name of the *interference* of the rays of light, were as simple and satisfactory as the principle itself is beautiful; but the verifications of it, drawn from the explanation it affords of phenomena apparently the most remote, are still more so. Newton's colours of thin films were the first phenomena to which its author applied it with full success. Its next remarkable application was to those of diffraction, of which, in the hands of M. Fresnel, a late eminent French geometer, it also furnished a complete explanation, and that, too, in cases to which Newton's hypothesis could not apparently be made to apply, and through a con plication of circumstances which might afford a very severe test of any hypothesis.

(289.) A simple and beautiful experiment on the interferences of polarized light due to Fresnel and Arago enabled them to bring Dr. Young's law to bear on the colours produced by crystallized plates in a polarized beam, and by so doing afforded a key to all the intricacies of these magnificent but complex phenomena. Nothing now was wanting to a rational theory of double refraction but to frame an hypothesis of some mode in which light might be conceived to be propagated through the elastic medium supposed to convey it in such a way as not to be contradictory to any of the facts, nor to the general laws of dynamics. This essential idea, without which every thing that had been before done would have been incomplete, was also furnished by Dr. Young, who, with a sagacity which would have done honour to Newton himself, had declared, that to accommodate the doctrine of Huyghens to the phenomena of polarized light it is necessary to conceive the mode of propagation of a luminous impulse through the ether, differently from that of a sonorous one through the air. In the latter, the particles of the air *advance* and *recede;* in the former, those of the ether must be supposed to *tremble laterally.*

(290.) Taking this as the groundwork of his reasoning, Fresnel succeeded in erecting on it a theory of polarization and double refraction, so happy in its adaptation to facts, and in the coincidence with experience of results deduced from it by the most intricate analysis, that it is difficult to conceive it unfounded. If it be so, it is at least the most curiously artificial system that science has yet witnessed; and whether it be so or not, so long as it serves to group together in one comprehensive point of view a mass of facts almost infinite in number and variety, to reason from one to another, and to establish analogies and relations between them; on whatever hypothesis it may be founded, or whatever arbitrary assumptions it may make respecting structures and modes of action, it can never be regarded as other than a most real and important accession to our knowledge.

(291.) Still, it is by no means impossible that the Newtonian theory of light, if cultivated with equal diligence with the Huyghenian, might lead to an equally plausible explanation of phenomena now regarded as beyond its reach. M. Biot is the author of the hypothesis we have already mentioned of a rotatory motion of the particles of light about their axes. He has employed it only for a very limited purpose; but it might doubtless be carried much farther; and by admitting only the regular emission of the luminous particles at equal intervals of time, and in similar states of motion from the shining body, which does not seem a very forced supposition, all the phenomena of interference at least would be readily enough explained without the admission of an ether.

(292.) The optical examination of crystallized substances affords one among many fine examples of the elucidation which every branch of science is capable of affording to every other. The indefatigable researches of Dr. Brewster and others have shown that the phenomena exhibited by polarized light in its transmission through crystals afford a certain indication of the most important points relating to the structure of the crystals themselves, and thus become most valuable characters by which to recognise their internal constitution. It was Newton who first showed of what importance as a physical character, — as the indication of other properties,—the action of a body on light might become; but the characters afforded by the use of polarized light as an instrument of experimental enquiry are sc marked and intimate, that they may almost be said to have furnished us with a kind of intellectual sense, by which we are enabled to scrutinize the internal arrangement of those wonderful structures which Nature builds up by her refined and invisible architecture, with a delicacy eluding our conception, yet with a symmetry and beauty which we are never weary of admiring. In this point of view the science of optics has rendered to mineralogy and crystallography services not less important than to astronomy by the invention of the telescope, or to natural history by that of the microscope; while the relations which have been discovered to exist between the optical properties of bodies and their crystalline forms, and even their chemical habitudes, have afforded numerous and beautiful instances of general laws concluded from laborious and painful induction, and curiously exemplifying the simplicity of nature as it emerges slowly from an entangled mass of particulars in which, at first, neither order nor connection can be traced.

CHAP. III.

OF COSMICAL PHENOMENA.

Astronomy and Celestial Mechanics.

(293.) ASTRONOMY, as has been observed in the former part of this discourse, as a science of observation, had made considerable progress among the ancients: indeed, it was the only branch of physical science which could be regarded as having been cultivated by them with any degree of assiduity or real success. The Chaldean and Egyptian records had furnished materials from which the motions of the sun and moon could be calculated with sufficient exactness for the prediction of eclipses; and some remarkable cycles, or periods of years in which the lunar eclipses return in very nearly the same order, had been ascertained by observation. Considering the extreme imperfection of their means of measuring time and space, this was, perhaps, as much as could have been expected at that early period, and it was followed up for a while in a philosophical spirit of just speculation, which, if continued, could hardly have failed to lead to sound and important conclusions.

(294.) Unfortunately, however, the philosophy of Aristotle laid it down as a principle, that the celestial motions were regulated by laws proper to themselves, and bearing no affinity to those which prevail on earth. By thus drawing a broad and impassable line of separation between celestial and terrestrial mechanics, it placed the former altogether out of the pale of experimental research, while it at the same time impeded the progress of the latter by the assumption of principles respecting natural and unnatural motions, hastily adopted from the most superficial and cursory remark, undeserving even the name of observation. Astronomy, therefore, continued for ages a science of mere record, in which theory had no part, except in so far as it attempted to conciliate the inequalities of the celestial motions with that assumed law of uniform circular revolution which was alone considered consistent with the perfection of the heavenly mechanism. Hence arose an unwieldy, if not self-contradictory, mass of hypothetical motions of sun, moon, and planets, in circles, whose centres were carried round in other circles, and these again in others without end,—"cycle on epicycle, orb on orb,"—till at length, as observation grew more exact, and fresh epicycles were continually added, the absurdity of so cumbrous a mechanism became too palpable to be borne. Doubts were expressed, to which the sarcasm of a monarch * gave a currency they might not have obtained in a period when men scarcely dared trust themselves to think; and at length Copernicus, promulgating his own, or reviving the Pythagorean doctrine, which places the sun in the centre of our system, gave to astronomy a simplicity which, contrasted with the complication of the preceding views, at once commanded assent.

* Alphonso of Castile, 1252.

(295.) An elegant writer *, whom we have before had occasion to quote, has briefly and neatly accounted for the confused notions which so long prevailed respecting the constitution of our system, and the difficulty experienced in acquiring a true notion of the disposition of its parts. "We see it," he observes, "not in *plan*, but in *section*." The reason of this is, that our point of observation lies in its general plane, but the notion we aim at forming of it is not that of its section, but of its plan. This is as if we should attempt to read a book, or make out the countries on a map, with the eye on a level with the paper. We can only judge directly of the distances of objects by their sizes, or rather of their change of distance by their change of size; neither have we any means of ascertaining, otherwise than indirectly, even their positions, one among the other, from their apparent places as seen by us. Now, the variations in apparent size of the sun and moon are too small to admit of exact measure without the use of the telescope, and the bodies of the planets cannot even be distinguished as having any distinct size with the naked eye.

(296.) The Copernican system once admitted, however, this difficulty of conception, at least, is effectually got over, and it becomes a mere problem of geometry and calculation to determine, from the observed places of a planet, its real orbit about the sun, and the other circumstances of its motion. This Kepler accomplished for the orbit of Mars, which he ascertained to be an ellipse having the sun in one of its foci; and the same law, being extended by inductive

* Jackson, Letters on Various Subjects, &c.

analogy to all the planets, was found to be verified in the case of each. This with the other remarkable laws which are usually cited in physical astronomy by the name of Kepler's laws, constitute undoubtedly the most important and beautiful system of geometrical relations which have ever been discovered by a mere inductive process, independent of any consideration of a theoretical kind. They comprise within them a compendium of the motions of all the planets, and enable us to assign their places in their orbits at any instant of time past or to come (disregarding their mutual perturbations), provided certain purely geometrical problems can be numerically resolved.

(297.) It was not, however, till long after Kepler's time that the real importance of these laws could be felt. Regarded in themselves, they offered, it is true, a fine example of regular and harmonious disposition in the greatest of all the works of creation, and a striking contrast to the cumbersome mechanism of the cycles and epicycles which preceded them; but there their utility seemed to terminate, and, indeed, Kepler was reproached, and not without a semblance of reason, with having rendered the actual calculation of the places of the planets more difficult than before, the resources of geometry being then inadequate to resolve the problems to which the strict application of his laws gave rise.

(298.) The first result of the invention of the telescope and its application to astronomical purposes, by Galileo, was the discovery of Jupiter's disc and satellites, — of a system offering a beautiful miniature of that greater one of which it forms a portion, and presenting to the eye of sense, at a single glance, that disposition of parts which in the planetary system itself is discerned only by the eye of reason and imagination (see 195.). Kepler had the satisfaction of seeing it ascertained, that the law which he had discovered to connect the times of revolution of the planets with their distances from the sun, holds good also when applied to the periods of circulation of these little attendants round the centre of their principal; thus demonstrating it to be something more than a mere empirical rule, and to depend on the intimate nature of planetary motion itself.

(299.) It had been objected to the doctrine of Copernicus, that, were it true, Venus should appear sometimes horned like the moon. To this he answered by admitting the conclusion, and averring that, should we ever be able to see its actual shape, it *would* appear so. It is easy to imagine with what force the application would strike every mind when the telescope confirmed this prediction, and showed the planet just as both the philosopher and his objectors had agreed it ought to appear. The history of science affords perhaps only one instance analogous to this. When Dr. Hutton expounded his theory of the consolidation of rocks by the application of heat, at a great depth below the bed of the ocean, and especially of that of marble by actual fusion; it was objected that, whatever might be the case with others, with calcareous or marble rocks, at least, it was impossible to grant such a cause of consolidation, since heat decomposes their substance and converts it into quicklime, by driving off the carbonic acid, and leaving a substance perfectly infusible, and incapable even of agglutination by heat. To this he replied, that the pressure under which the heat was applied would prevent the escape of the carbonic acid; and that being retained, it might be expected to give that fusibility to the compound which the simple quicklime wanted. The next generation saw this anticipation converted into an observed fact, and verified by the direct experiments of Sir James Hall, who actually succeeded in melting marble, by retaining its carbonic acid under violent pressure.

(300.) Kepler, among a number of vague and even wild speculations on the causes of the motions whose laws he had developed so beautifully and with so much patient labour, had obtained a glimpse of the general law of the inertia of matter, as applicable to the great masses of the heavenly bodies as well as to those with which we are conversant on the earth. After Kepler, Galileo, while he gave the finishing blow to the Aristotelian dogmas which erected a barrier between the laws of celestial and terrestrial motion, by his powerful argument and caustic ridicule, contributed, by his investigations of the laws of falling bodies and the motions of projectiles, to lay the foundation of a true system of dynamics, by which motions could be determined from a knowledge of the forces producing them, and forces from the motions they produce. Hooke went yet farther, and obtained a view so distinct of the mode in which the planets might be retained in their orbits by the sun's attraction, that, had his mathematical attainments been equal to his philosophical acumen, and his scientific pursuits been less various and desultory, it can hardly be doubted that he would have arrived at a knowledge of the law of gravitation.

(301.) But every thing which had been done towards this great end, before Newton, could only be regarded as smoothing some first obstacles, and preparing a state of knowledge, in which powers like his could be effectually exerted. His wonderful combination of mathematical skill with physical research enabled him to invent, at pleasure, new and unheard-of methods of investigating the effects of those causes which his clear and penetrating mind detected in operation. Whatever department of science he touched, he may be said to have formed afresh. Ascending by a series of close-compacted inductive arguments to the highest axioms of dynamical science, he succeeded in applying them to the complete explanation of all the great astronomical phenomena, and many of the minuter and more enigmatical ones. In doing this, he had every thing to create: the mathematics of his age proved totally inadequate to grapple with the numerous difficulties which were to be overcome; but this, so far from discouraging him, served only to afford new opportunities for the exertion of his genius, which, in the invention of the method of fluxions, or, as it is now more generally called, the differential calculus, has supplied a means of discovery, bearing the same proportion to the methods previously in use, that the steam-engine does to the mechanical powers employed before its invention. Of the optical discoveries of Newton we have already spoken; and if the magnitude of the objects of his astronomical discoveries excite our admiration of the mental powers which could so familiarly grasp them, the minuteness of the researches into which he there set the first example of entering, is no less calculated to produce a corresponding impression. Whichever way we turn our view, we find ourselves compelled to bow before his genius, and to assign to the name of NEWTON a place in our veneration which belongs to no other in the annals of science. His era marks the accomplished maturity of the human reason as applied to such objects. Every thing which went before might be more properly compared to the first imperfect attempts of childhood, or the essays of inexpert, though promising, adolescence. Whatever has been since performed, however great in itself, and worthy of so splendid and auspicious a beginning, has never, in point of intellectual effort, surpassed that astonishing one which produced the Principia.

(302.) In this great work, Newton shows all the celestial motions known in his time to be consequences of the simple law, that every particle of matter attracts every other particle in the universe with a force proportional to the product of their masses directly, and the square of their mutual distance inversely, and is itself attracted with an equal force. Setting out from this, he explains how an attraction arises between the great spherical masses of which our system consists, regulated by a law precisely similar in its expression; how the elliptic motions of planets about the sun, and of satellites about their primaries, according to the exact rules inductively arrived at by Kepler, result as necessary consequences from the same general law of force; and how the orbits of comets themselves are only particular cases of planetary movements. Thence proceeding to applications of greater difficulty, he explains how the perplexing inequalities of the moon's motion result from the sun's disturbing action; how tides arise from the unequal attraction of the sun as well as of the moon on the earth, and the ocean which surrounds it; and, lastly, how the precession of the equinoxes is a necessary consequence of the very same law.

(303.) The immediate successors of Newton found full occupation in verifying his discoveries, and in extending and improving the mathematical methods which it had now become manifest were to prove the keys to an inexhaustible treasure of knowledge. The simultaneous but independent discovery of a method of mathematical investigation in every respect similar to that of Newton, by Leibnitz, while it created a degree of national jealousy which can now only be pitied, had the effect of stimulating the continental geometers to its cultivation, and impressing on it a character more entirely independent of the ancient geometry, to which Newton was peculiarly attached. It was fortunate for science that it did so; for it was speedily found that (with one fine exception on the part of our countryman Maclaurin, followed up, after a long interval, by the late Professor Robison of Edinburgh, with equal elegance,) the geometry of Newton was like the bow of Ulysses, which none but its master could bend; and that, to render his methods available beyond the points to which he himself carried them, it was necessary to strip them of every vestige of that antique dress in which he had delighted to clothe them. This, however, the countrymen of Newton were very unwilling to do; and they paid the penalty in finding themselves condemned to the situation of lookers on, while their continental neighbours both in Germany and France were pushing forward in the career of mathematico-physical discovery with emulous rapidity.

(304.) The legacy of research which Newton may be said to have left to his successors was truly immense. To pursue, through all its intricacies, the consequences of the law of gravitation; to account for all the inequalities of the planetary movements, and the infinitely more complicated, and to us more important ones, of the moon; and to give, what Newton himself certainly never entertained a conception of, a demonstration of the stability and permanence of the system, under all the accumulating influence of its internal perturbations; this labour, and this triumph, were reserved for the succeeding age, and have been shared in succession by Clairaut, D'Alembert, Euler, Lagrange and Laplace. Yet so extensive is the subject, and so difficult and intricate the purely mathematical enquiries to which it leads, that another century may yet be required to go through with the task. The recent discoveries of astronomers have supplied matter for investigation, to the geometers of this and the next generation, of a difficulty far surpassing any thing that had before occurred. Five primary planets have been added to our system; four of them since the commencement of the present century, and these, singularly deviating from the general analogy of the others, and offering *cases of difficulty* in theory, which no one had before contemplated. Yet even the intricate questions to which these bodies have given rise seem likely to be surpassed by those which have come into view, with the discovery of several comets revolving in elliptic orbits, like the planets, round the sun, in very moderate periods. But the resources of modern geometry seem, so far from being exhausted, to increase with the difficulties they have to encounter, and already, among the successors of Lagrange and Laplace, the present generation has to enumerate a powerful array of names, which promise to render it not less celebrated in the annals of physico-mathematical research than that which has just passed away.

(305.) Meanwhile the positions, figures, and dimensions of all the planetary orbits, are now well known, and their variations from century to century in great measure determined; and it has been generally demonstrated, that all the changes which the mutual actions of the planets on each other can produce in the course of indefinite ages, are *periodical*, that is to say, increasing to a certain extent (and that never a very great one), and then again decreasing; so that the system can never be destroyed or subverted by the mutual action of its parts, but keeps constantly oscillating, as it were, round a certain mean state, from which it can never deviate to any ruinous extent. In particular, the researches of Laplace and Lagrange have demonstrated the absolute invariability of the mean distance of each planet from the sun, and consequently of its periodic time. Relying on these grand discoveries, we are enabled to look forward, from the point of time which we now occupy, many thousands of years into futurity, and predict the state of our system without fear of material error, but such as may arise from causes whose existence at present we have no reason to suppose, or from interference which we have no right to anticipate.

(306.) A correct enumeration and description of the fixed stars in catalogues, and an exact knowledge of their position, supply the only effectual means we can have of ascertaining what changes they are liable to, and what motions, too slow to deprive them of their usual epithet, *fixed*, yet sufficient to produce a sensible change in the lapse of ages, may exist among them. Previous to the invention of the compass, they served as guides to the navigator by night; but for this purpose, a very moderate knowledge of a few of the principal ones sufficed. Hipparchus was the first astronomer, who, excited by the appearance of a new star, conceived the idea of forming a catalogue of the stars, with a view to its use as an astronomical record, "by which," says Pliny, "posterity will be able to discover, not only whether they are born and die, but also whether they change their places, and whether they increase or decrease." His catalogue, containing more than 1000 stars, was constructed about 128 years before Christ. It was in the course of the laborious discussion of his own and former observations of them, undertaken with a view to the formation of this catalogue, that he first recognised the fact of that slow, general advance of all the stars eastward, when compared with the place of the equinox, which is known under the name of the precession of the equinoxes, and which Newton succeeded in referring to a motion in the earth's axis, produced by the attraction of the sun and moon.

(307.) Since Hipparchus, at various periods in the history of astronomy, catalogues of stars have been formed, among which that of Ulugh Begh, comprising about 1000 stars, constructed in 1437, is remarkable as the production of a sovereign prince, working personally in conjunction with his astronomers; and that of Tycho Brahe, containing 777 stars, constructed in 1600, as having originated in a phenomenon similar to that which drew the attention of Hipparchus. In more recent times, astronomers provided with the finest instruments their respective eras could supply, and established in observatories, munificently endowed by the sovereigns and governments of different European nations, have vied and are still vying with each other, in extending the number of registered stars, and giving the utmost possible degree of accuracy to the determination of their places. Among these, it would be ungrateful not to claim especial notice for the superb series of observations which, under a succession of indefatigable and meritorious astronomers, has, for a very long period, continued to emanate from our own national observatory of Greenwich.

(308.) The distance of the fixed stars is so immense, that every attempt to assign a limit, *within which* it *must* fall, has hitherto failed The enquiries of astronomers of all ages have been directed to ascertain this distance, by taking the dimensions of our own particular system of sun and planets, or of the earth itself, as the unit of a scale on which it might be measured. But although many have imagined that their observations afforded grounds for the decision of this interesting point, it has uniformly happened either that the phenomena on which they relied have proved to be referable to other causes not previously known, and which the superior accuracy of their researches has for the first time brought to light; or to errors arising from instrumental imperfections and unavoidable defects of the observations themselves.

(309.) The only indication we can expect to obtain of the actual distance of a star, would consist in an annual change in its apparent place corresponding to the motion of the earth round the sun, called its *annual parallax*, and which is nothing more than the measure of the apparent size of the earth's orbit as seen from the star. Many observers have thought they have detected a measurable amount of this parallax; but as astronomical instruments have advanced in perfection, the quantity which they have successively assigned to it has been continually reduced within narrower and narrower limits, and has invariably been commensurate with the errors to which the instruments used might fairly be considered liable. The conclusion this strongly presses on us is, that it is really a quantity too small to admit of distinct measurement in the present state of our means for that purpose; and that, therefore, the distance of the stars must be a magnitude of such an order as the imagination almost shrinks from contemplating. But this increase in our scale of dimension calls for a corresponding enlargement of conception in all other respects. The same reasoning which places the stars at such immeasurable remoteness, exalts them at the same time into glorious bodies, similar to, and even far surpassing, our own sun, the centres perhaps of other planetary systems, or fulfilling purposes of which we can have no idea, from any analogy in what passes immediately around us.

(310.) The comparison of catalogues, published at different periods, has given occasion to many curious remarks, respecting changes both of place and brightness among the stars, to the discovery of variable ones which lose and recover their lustre periodically, and to that of the disappearance of several from the heavens so completely as to have left no vestige discernible even by powerful telescopes. In proportion as the construction of astronomical and optical instruments has gone on improving, our knowledge of the contents of the heavens has undergone a corresponding extension, and, at the same time, attained a degree of precision which could not have been anticipated in former ages. The places of all the principal stars in the northern hemisphere, and of a great many in the southern, are now known to a degree of nicety which must infallibly detect any real motions that may exist among them, and has in fact done so, in a great many instances, some of them very remarkable ones.

(311.) It is only since a comparatively recent date, however, that any great attention has been bestowed on the smaller stars, among which there can be no doubt of the most interesting and instructive phenomena being sooner or later brought to light. The minute examination of them with powerful telescopes, and with delicate instruments for the determination of their places, has, indeed, already produced immense catalogues and masses of observations, in which thousands of stars invisible to the naked eye are registered; and has led to the discovery of innumerable important and curious facts, and disclosed the existence of whole classes of celestial objects, of a nature so wonderful as to give room for unbounded speculation on the extent and construction of the universe.

(312.) Among these, perhaps the most remarkable are the revolving double stars, or stars which, to the naked eye or to inferior telescopes, appear single; but, if examined with high magnifying powers, are found to consist of two individuals placed almost close together, and which, when carefully watched, are (many of them) found to revolve in regular elliptic orbits about each other; and so far as we have yet been able to ascertain, to obey the same laws which regulate the planetary movements. There is nothing calculated to give a grander idea of the scale on which the sidereal heavens are constructed than these beautiful systems. When we see such magnificent bodies united in pairs, undoubtedly by the same bond of mutual gravitation which holds together our own system, and sweeping over their enormous orbits, in periods comprehending many centuries, we admit at once that they must be accomplishing ends in creation which will remain for ever unknown to man; and that we have here attained a point in science where the human intellect is compelled to acknowledge its weakness, and to feel that no conception the wildest imagination can form will bear the least comparison with the intrinsic greatness of the subject.

Geology.

(313.) The researches of physical astronomy are confessedly incompetent to carry us back to the origin of our system, or to a period when its state was, in any great essential, different from what it is at present. So far as the causes now in action go, and so far as our calculations will enable us to estimate their effects, we are equally unable to perceive in the general phenomena of the planetary system either the evidence of a beginning, or the prospect of an end. Geometers, as already stated, have demonstrated that, in the midst of all the fluctuations which can possibly take place in the elements of the orbits of the planets, by reason of their mutual attraction, the general balance of the parts of the system will always be preserved, and every departure from a mean state periodically compensated. But neither the researches of the physical astronomer, nor those of the geologist, give us any ground for regarding our system, or the globe we inhabit, as of eternal duration. On the contrary, there are circumstances in the physical constitution of our own planet which at least obscurely point to an origin and a formation, however remote, since it has been found that the figure of the earth is not globular but elliptical, and that its attraction is such as requires us to admit the interior to be more dense than the exterior, and the density to increase with some degree of regularity from the surface towards the centre, and *that*, in layers arranged elliptically round the centre, circumstances which could scarcely happen without some such successive deposition of materials as would enable pressure to be propagated with a certain degree of freedom from one part of the mass to another, even if we should hesitate to admit a state of primitive fluidity.

(314.) But from such indications nothing distinct can be concluded; and if we would speculate to any purpose on a former state of our globe and on the succession of events which from time to time may have changed the condition and form of its surface, we must confine our views within limits far more restricted, and to subjects much more within the reach of our capacity, than either the creation of the world or its assumption of its present figure. These, indeed, were favourite speculations with a race of geologists now extinct; but the science itself has undergone a total change of character, even within the last half century, and is brought, at length, effectually within the list of the inductive sciences. Geologists now no longer bewilder their imaginations with wild theories of the formation of the globe from chaos, or its passage through a series of hypothetical transformations, but rather aim at a careful and accurate examination of the records of its former state, which they find indelibly impressed on the great features of its actual surface, and to the evidences of former life and habitation which organised remains imbedded and preserved in its strata indisputably afford.

(315.) Records of this kind are neither few nor vague; and though the obsoleteness of their language when we endeavour to interpret it too minutely, may, and no doubt often does, lead to misapprehension, still its general meaning is, on the whole, unequivocal and satisfactory. Such records teach us, in terms too plain to be misunderstood, that the whole or nearly the whole of our present lands and continents were formerly at the bottom of the sea, where they received deposits of materials from the wearing and degradation of other lands not now existing, and furnished receptacles for the remains of marine animals and plants inhabiting the ocean above them, as well as for similar spoils of the land washed down into its bosom.

(316.) These remains are occasionally brought to light; and their examination has afforded indubitable evidence of the former existence of a state of animated nature widely different from what now obtains on the globe, and of a period anterior to that in which it has been the habitation of man, or rather, indeed, of a series of periods, of unknown duration, in which both land and sea teemed with forms of animal and vegetable life, which have successively disappeared and given place to others, and these again to new races approximating gradually more and more nearly to those which now inhabit them, and at length comprehending species which have their counterparts existing.

(317.) These wrecks of a former state of nature,

thus wonderfully preserved (like ancient medals and inscriptions in the ruins of an empire), afford a sort of rude chronology, by whose aid the successive depositions of the strata in which they are found may be marked out in epochs more or less definitely terminated, and each characterized by some peculiarity which enables us to recognise the deposits of any period, in whatever part of the world they may be found. And, so far as has been hitherto investigated, the *order* of succession in which these deposits have been formed appears to have been the same in every part of the globe.

(318.) Many of the strata which thus bear evident marks of having been deposited at the bottom of the sea, and of course in a horizontal state, are now found in a position highly inclined to the horizon, and even occasionally vertical. And they often bear evident marks of violence, in their bending and fracture, in the dislocation of parts which were once contiguous, and in the existence of vast collections of broken fragments which afford every proof of great violence having been used in accomplishing some at least of the changes which have taken place.

(319.) Besides the rocks which carry this internal evidence of submarine deposition, are many which exhibit no such proofs, but on the contrary hold out every appearance of owing their origin to volcanoes or to some other mode of igneous action; and in every part of the world, and among strata of all ages, there occur evidences of such action so abundant, and on such a scale, as to point out the volcano and the earthquake as agents which may have been instrumental in the production of those changes of level, and those violent dislocations which we perceive to have taken place.

(320.) At all events, in accounting for those changes, geologists have no longer recourse, as formerly, to causes purely hypothetical, such as a shifting of the earth's axis of rotation, bringing the sea to overflow the land, by a change in the place of the longer and shorter diameters of the spheroidal figure, nor to tides produced by the attraction of comets suddenly approaching very near the earth, nor to any other fanciful and arbitrarily assumed hypotheses; but rather endeavour to confine themselves to a careful consideration of causes evidently in action at present, with a view to ascertain how far they, in the first instance, are capable of accounting for the facts observed, and thus legitimately bringing into view, as residual phenomena, those effects which cannot be so accounted for. When this shall have been in some measure accomplished, we shall be able to pronounce with greater security than at present respecting the necessity of admitting a long succession of tremendous and ravaging catastrophes and cataclysms, — epochs of terrific confusion and violence which many geologists (perhaps with justice) regard as indispensable to the explanation of the existing features of the world. We shall learn to distinguish between the effects which require for their production the sudden application of convulsive and fracturing efforts, and those, probably not less extensive, changes which may have been produced by forces equally or more powerful, but acting with less irregularity, and so distributed over time as to produce none of those *interregnums* of chaotic anarchy which we are apt to think (perhaps erroneously) great disfigurements of an order so beautiful and harmonious as that of nature.

(321.) But to estimate justly the effects of causes now in action in geology is no easy task. There is no *à priori* or deductive process by which we can estimate the amount of the annual erosion, for instance, of a continent by the action of meteoric agents, rain, wind, frost, &c., nor the quantity of destruction produced on its coasts by the direct violence of the sea, nor the quantity of lava thrown up *per annum* by volcanoes over the whole surface of the earth, nor any similar effect. And to consult experience on all such points is a slow and painful process if rightly gone into, and a very fallible one if only partially executed. Much, then, at present must be left to opinion, and to that sort of clear-judging tact which sometimes anticipates experience; but this ought not to stand in the way of our making every possible effort to obtain accurate information on such points, by which alone geology can be rendered, if not an experimental science, at least a science of that kind of active observation which forms the nearest approach to it, where actual experiment is impossible.

(322.) Let us take, for example, the question, "What is the actual direction in which changes of relative level are taking place between the existing continents and seas?" If we consult partial experience, that is, *all* the information that we possess respecting ancient sea-marks, soundings, &c., we shall only find ourselves bewildered in a mass of conflicting, because imperfect, evidence. It is obvious that the only way to decide the point is to ascertain, by very precise and careful observations at proper stations on coasts, selected at points where there exist natural marks not liable to change in the course of at least a century, the true elevation of such marks above the *mean* level of the sea, and to multiply these stations sufficiently over the whole globe to be capable of affording real available knowledge. Now, this is not a very easy operation (considering the accuracy required); for the *mean* level of the sea can be determined by no single observation, any more than the mean height of the barometer at a given station, being affected both by periodical and accidental fluctuations due to tides, winds, waves, and currents. Yet if an instrument adapted for the purpose were constructed, and rendered easily attainable, and rules for its use carefully drawn up, there is little doubt we should soon (by the industry of observers scattered over the world) be in possession of a most valuable mass of information, which could not fail to afford a point of departure for the next generation, and furnish ground for the only kind of argument which ever can be conclusive on such subjects.

(323.) Geology, in the magnitude and sublimity of the objects of which it treats, undoubtedly ranks, in the scale of the sciences, next to astronomy; like astronomy, too, its progress depends on the continual accumulation of observations carried on for ages. But, unlike astronomy, the observations on which it depends, when the whole extent of the subject to be explored is taken into consideration, can hardly yet be said to be more than commenced. Yet, to make up for this, there is one important difference, that while in the latter science it is impossible to recall the past or anticipate the future, and observation is in consequence limited to a single fact in a single moment; in the former, the records of the past are always present; — they may be examined and re-examined as often as we please, and require nothing but diligence and judgment to put us in possession of their whole contents. Only a very small part of the surface of our globe has, however, been accurately examined in detail, and of that small portion we are only able to scratch the mere exterior, for so we must consider those excavations which we are apt to regard as searching the bowels of the earth; since the deepest mines which have been sunk penetrate to a depth hardly surpassing the ten thousandth part of the distance between its surface and its centre. Of course inductions founded on such limited examination can only be regarded as provisional, except in those remarkable cases where the same great formations in the same order have been recognised in very distant quarters, and without exception. This, however, cannot long be the case. The spirit with which the subject has been prosecuted for many years in our own country has been rewarded with so rich a harvest of surprising and unexpected discoveries, and has carried the investigation of our island into such detail, as to have excited a corresponding spirit among our continental neighbours; while the same zeal which animates our countrymen on their native shore accompanies them in their sojourns abroad, and has already begun to supply a fund of information respecting the geology of our Indian possessions, as well as of every other point where English intellect and research can penetrate.

(324.) Nothing can be more desirable than that every possible facility and encouragement should be afforded for such researches, and indeed to the pursuits of the enlightened resident or traveller in every department of science, by the representatives of our national authority wherever our power extends. By these only can our knowledge of the actual state of the surface of the globe, and that of the animals and vegetables of the ancient continents and seas, be extended and perfected, while more complete information than we at present possess of the habits of those actually existing, and the influence of changes of climate, food, and circumstances, on them, may be expected to render material assistance to our speculations respecting those which have become extinct.

CHAP. IV.

OF THE EXAMINATION OF THE MATERIAL CONSTITUENTS OF THE WORLD.

Mineralogy.

(325.) THE consideration of the history and structure of our globe, and the examination of the fossil contents of its strata, lead us naturally to consider the materials of which it consists. The history of these materials, their properties as objects of philosophical enquiry, and their application to the useful arts and the embellishments of life, with the characters by which they can be certainly distinguished one from another, form the object of mineralogy, taken in its most extended sense.

(326.) There is no branch of science which presents so many points of contact with other departments of physical research, and serves as a connecting link between so many distant points of philosophical speculation, as this. To the geologist, the chemist, the optician, the crystallographer, the physician, it offers especially the very elements of their knowledge, and a field for many of their most curious and important enquiries. Nor, with the exception of chemistry, is there any which has undergone more revolutions, or been exhibited in a greater variety of forms. To the ancients it could scarcely be said to be at all known, and up to a comparatively recent period, nothing could be more imperfect than its descriptions, or more inartificial and unnatural than its classification. The more important minerals in the arts, indeed, those used for economical purposes and those from which metals were extracted, had a certain degree of attention paid to them, for the sake of their utility and commercial value, and the precious stones for that of ornament. But until their crystalline forms were attentively observed and shown to be determinate characters on which dependence could be placed, no mineralogist could give any correct account of the real distinction between one mineral and another.

(327.) It was only, however, when chemical analysis had acquired a certain degree of precision and universal applicability that the importance of mineralogy as a science began to be recognised, and the connection between the external characters of a stone and its ingredient constituents brought into distinct notice. Among these characters, however, none were found to possess that eminent distinctness which the crystalline form offers; a character, in the highest degree geometrical, and affording, as might be naturally supposed, the strongest evidence of its necessary connection with the intimate constitution of the substance. The full importance of this character was, however, not felt until its connection with the texture or cleavage of a mineral was pointed out, and even then it required numerous and striking instances of the critical discernment of Haüy and other eminent mineralogists in predicting from the measurements of the angles of crystals which had been confounded together that differences would be found to exist in their chemical composition, all which proved fully justified in their result before the essential value of this character was acknowledged. This was no doubt in great measure owing to the high importance set by the German mineralogists on those external characters of touch, sight, weight, colour, and other sensible qualities, which are little susceptible, with the exception of weight, of exact determination, and which are subject to material variations in different specimens of the same mineral. By degrees, however, the necessity of ascribing great weight to a character so definite was admitted, especially when it was considered that the same step which pointed out the intimate connection of external form with internal structure furnished the mineralogist with the means of reducing all the forms of which a mineral is susceptible under one general type, or primitive form, and afforded grounds for an elegant theoretical account of the assumption of definite figures *ab initio*.

(328.) A simple and elegant invention of Dr. Wollaston, the reflecting goniometer, gave a fresh impulse to that view of mineralogy which makes the crystalline form the essential or leading character, by putting it in the power of every one, by the examination of even the smallest portion of a broken crystal, to ascertain and verify that essential character on which the identity of a mineral in the system of Haüy was made to depend. The application of so ready and exact a method speedily led to important results, and to a still nicer discrimination of mineral species than could before be attained; and the confirmation given to these results by chemical analysis stamped them with a scientific and decided character which they have retained ever since.

(329.) Meanwhile the progress made in chemical analysis had led to the important conclusion that every chemical compound susceptible of assuming the solid state assumed with it a determinate crystalline form; and the progress of optical science had shown that the fundamental crystalline form, in the case at least of transparent bodies, drew with it a series of optical properties no less curious than important in relation to the affections of light in its passage through such substances. Thus, in every point of view, additional importance became added to this character; and the study of the crystalline forms of bodies in general assumed the form of a separate and independent branch of science, of which the geometrical forms of the mineral world constituted only a particular case. Mineralogy, however, as a branch of natural history, remains still distinct either from optics or crystallography. The mineralogist is content, and thinks he has performed his task, if not as a natural historian at least as a classifier and arranger, if he only gives such a characteristic description of a mineral as shall effectually distinguish it from every other, and shall enable any one who may encounter such a body in any part of the world to impose on it its name, assign it a place in his system, and turn to his books for a further description of all that the chemist, the optician, the lapidary, or the artist, may require to know. Still this is no easy matter: the laborious researches of the most eminent mineralogists can hardly yet be said to have effectually accomplished it; and its difficulty may be appreciated by the small number of simple minerals, or minerals of perfectly definite and well-marked characters, which have been hitherto made out. Nor can this indeed be wondered at, when we consider that by far the greater portion of the rocks and stones which compose the external crust of the globe consists of nothing more than the accumulated *detritus* of older rocks, in which the fragments and powder of an infinite variety of substances are mingled together, in all sorts of varying proportions, and in such a way as to defy separation. Many of these rocks, however, so compounded, occur with sufficient frequency and uniformity of character to have acquired names and to have been usefully applied; indeed, in the latter respect, minerals of this description far surpass all the others. As objects of natural history, therefore, they are well worthy of attention, however difficult it may be to assign them a place in any artificial arrangement.

(330.) This paucity of simple minerals, however, is probably rather apparent than real, and in proportion as the researches of the chemist and crystallographer shall be extended throughout nature, they will no doubt become much more numerous. Indeed, in the great laboratories of nature it can hardly be doubted that almost every kind of chemical process is going forwards, by which compounds of every description are continually forming. Accordingly, it is remarked, that the lavas and ejected scoriæ of volcanoes are receptacles in which mineral products previously unknown are constantly discovered, and that the primitive formations, as they are called in geology, which bear no marks of having been produced by the destruction of others, are also remarkable for the beauty and distinctness of character of their minerals.

(331.) The great difficulty which has been experienced in attempts to classify mineral substances by their chemical constituents has arisen from the observed presence, in some specimens of minerals bearing that general resemblance in other respects as well as agreement in form which would seem to entitle them to be considered as alike, of ingredients foreign to the usual composition of the species, and that occasionally in so large a proportion as to render it unjustifiable to refer their occurrence to accidental impurities. These cases, as well as some anomalies observed in the classification of minerals by their crystalline forms, which seemed to show that the same substance might occasionally appear under two distinct forms, as well as some remarkable coincidences between the forms of substances quite distinct from each other in a chemical point of view, have within a recent period given rise to a branch of the science of crystallography of a very curious and important nature. The *isomorphism* of certain groups of chemical elements has already afforded us an example illustrative of the manner in which inductions sometimes receive unexpected verifications (see 180.). The laws and relations thus brought to light are among the most curious and interesting parts of modern science, and seem likely in their further developement to afford ample scope for the exercise of chemical and mineralogical research. They have already afforded innumerable fine examples of that important step in science by which anomalies disappear, and occasional incongruities become reconciled under more general expressions of physical laws, and thus unite in affording support to those very views which they promised, when first observed, to overset. Nothing, indeed, can be more striking than to see the very ingredient which every previous chemist and mineralogist would agree to disregard and reject as a mere casual impurity brought forward and appealed to in support of a theory expressly directed to the object of rescuing science from the imputation of disregarding, under any circumstances, the plain results of direct experiment.

Chemistry.

(332.) The laws which concern the intimate constitution of bodies, not as respects their *structure* or the manner in which their parts are put together, but as regards their *materials* or the ingredients of which those parts are composed, form the objects of chemistry. A solid body may be regarded as a fabric, more or less regularly and artificially constructed, in which the materials and the workmanship may be separately considered, and in which, though the latter be ruined and confounded by violence, the former remain unchanged in their nature, though differently arranged. In liquid or aërial bodies, too, though there prevails a less degree of difference in point of structure, and a greater facility of dispersion and dissipation, than in solids, yet an equal diversity of *materials* subsists, giving to them properties differing extremely from each other.

(333.) The inherent activity of matter is proved not only by the production of motion by the mutual attractions and repulsions of distant or contiguous masses, but by the changes and apparent transformations which different substances undergo in their sensible qualities by mere mixture. If water be added to water, or salt to salt, the effect is an increase of quantity, but no change of quality. In this case, the mutual action of the particles is entirely mechanical. Again, if a blue powder and a yellow one, each perfectly dry, be mixed and well shaken together, a green powder will be produced; but this is a mere effect arising in the eye from the intimate mixture of the yellow and blue light separately and independently reflected from the minute particles of each; and the proof is had by examining the mixture with a microscope, when the yellow and blue grains will be seen separate and each quite unaltered. If the same experiment be tried with coloured liquids, which are susceptible of mixing without chemical action, a compound colour is likewise produced, but no examination with magnifiers is in that case sufficient to detect the ingredients; the reason obviously being, the excessive minuteness of the parts, and their perfect intermixture, produced by agitating two liquids together. From the mixture of two powders, extreme patience would enable any one, by picking out with a magnifier grain after grain, to separate the ingredients. But when liquids are mixed, no mechanical separation is any longer practicable; the particles are so minute as to elude all search. Yet this does not hinder us from regarding such a compound as still a mere mixture, and its properties are accordingly intermediate between those of the liquids mixed. But this is far from being the case with all liquids. When a solution of potash, for example, and another of tartaric acid, each perfectly liquid, are mixed together in proper proportions, a great quantity of a solid saline substance falls to the bottom of the containing vessel, which is quite different from either potash or tartaric acid, and the liquid from which it subsided offers no indications by its taste or other sensible qualities of the ingredients mixed, but of something totally different from either. It is evident that this is a phenomenon widely different from that of mere mixture; there has taken place a great and radical change in the intimate nature of the ingredients, by which a new substance is produced which had no existence before. And it has been produced by the *union* of the ingredients presented to each other; for when examined it is found that nothing has been *lost*, the weight of the whole mixture being the sum of the weights mixed. Yet the potash and tartaric acid have disappeared entirely, and the weight of the new product is found to be exactly equal to that of the tartaric acid and potash employed, taken together, abating a small portion held in solution in the liquid, which may be obtained however by evaporation. They have therefore combined, and adhere to one another with a cohesive force sufficient to form a solid out of a liquid; a force which has thus been called into action by merely presenting them to each other in a state of solution.

(334.) It is the business of chemistry to investigate these and similar changes, or the reverse of such changes, where a single substance is resolved into two or more others, having different properties from it, and from each other, and to enquire into all the circumstances which can influence them; and either determine, modify, or suspend their accomplishment, whether such influence be exercised by heat or cold, by time and rest, or by agitation or pressure, or by any of those agents of which we have acquired a knowledge, such as electricity, light, magnetism, &c.

(335.) The wonderful and sudden transformations with which chemistry is conversant, the violent activity often assumed by substances usually considered the most inert and sluggish, and, above all, the insight it gives into the nature of innumerable operations which we see daily carried on around us, have contributed to render it the most popular, as it is one of the most extensively useful, of the sciences; and we shall, accordingly, find none which have sprung forward, during the last century, with such extraordinary vigour, and have had such extensive influence in promoting corresponding progress in others. One of the chief causes of its popularity is, perhaps, to be sought for in this, that it is, of all the sciences, perhaps, the most completely an experimental one; and even its theories are, for the most part, of that generally intelligible and readily applicable kind, which demand no intense concentration of thought, and lead to no profound mathematical researches. The simple process of inductive generalization, grounded on the examination of numerous facts, all of them presenting considerable intrinsic interest, has sufficed, in most instances, to lead, by a clear and direct road, to its highest laws yet known. But, on the other hand, these laws, when stated, are not yet fully sufficient to lead us, except in very limited cases, to a deductive knowledge of particulars never before examined, at least, not without great caution, and constant appeal to experiment as a check on our reasoning; so that we are justified in regarding the *axioms* of chemistry, the true handles of deductive reasoning, as still unknown, and, perhaps, likely long to remain so. This is no fault of its cultivators, who have comprised in their list the highest and most varied talents and industry, but of the inherent complexity of the subject, and the infinite multitude of causes which are concerned in the production of every, even the simplest, chemical phenomenon.

(336.) The history of chemistry (on which, however, we are not about to enlarge,) is one of great interest to those who delight to trace the steps by which mankind advance to the discovery of truth through a series of mistakes and failures. It may be divided, 1st, into the period of the alchemists, a lamentable epoch in the annals of intellectual wandering; 2dly, that of the phlogistic doctrines of Beccher and Stahl, in which, as if to prove the perversity of the human mind, of two possible roads the wrong was chosen; and a theory obtained universal credence on the credit of great names, ingenious views, and loose experiments, which is negatived, *in every instance*, by an appeal to the balance. This, too, happened, not by reason of unlucky coincidences, or individual oversights, but of necessity, and from an inherent defect of the theory itself, which thus impeded the progress of the science, as far as a science of experiment can be impeded by a false theory, by perplexing its cultivators with the appearance of contradictions in their experiments where none really subsisted, by destroying all their confidence in the numerical exactness of their own results, and by involving the subject in a mist of visionary and hypothetical causes in place of the true acting principles. Thus, in the combustion of any substance which is incapable of flying away in fumes, an increase of weight takes place,—the ashes are heavier than the fuel. Whenever this was observed, however, it was passed carelessly over as arising from the escape of phlogiston, or the principle of inflammability, which was considered as being either the element of fire itself, or in some way combined with it, and thus essentially *light*. It is now known that the increase of weight is owing to the absorption of, and combination with, a quantity of a peculiar ingredient called *oxygen*, from the air, a principle essentially *heavy*. So far as weight is concerned, it makes no difference whether a body having weight enters, or one having levity escapes; but there is this plain difference in a philosophical point of view, that oxygen is a real producible substance, and phlogiston is no such thing: the former is a *vera causa*, the latter an hypothetical being, introduced to account for what the other accounts for much better.

(337.) The third age of chemistry—that which may be called emphatically modern chemistry—

commenced (in 1786) when Lavoisier, by a series of memorable experiments, extinguished for ever this error, and placed chemistry in the rank of one of the exact sciences,—a science of number, weight, and measure. From that epoch to the present day it has constantly advanced with an accelerated progress, and at this moment may be regarded as more progressive than ever. The principal features in this progress may be comprised under the following general heads:—

1. The discovery of the proximate, if not the ultimate, elements of all bodies, and the enlargement of the list of known elements to its present extent of between fifty and sixty substances.
2. The developement of the doctrine of latent heat by Black, with its train of important consequences, including the scientific theory of the steam-engine.
3. The establishment of Wenzel's law of definite proportions on his own experiments, and those of Richter, a discovery subsequently merged in the greater generality of the atomic theory of Dalton.
4. The precise determination of the atomic weights of the different chemical elements, mainly due to the astonishing industry of Berzelius, and his unrivalled command of chemical resources, as well as to the researches of the other chemists of the Swedish and German school, and of our countryman, Dr. Thomson.
5. The assimilation of gases and vapours, by which

we are led to regard the former, universally, as particular cases of the latter, a generalization resulting chiefly from the experiments of Faraday on the condensation of the gases, and those of Gay Lussac and Dalton, on the laws of their expansion by heat compared with that of vapours.

6. The establishment of the laws of the combination of gases and vapours by definite volumes, by Gay Lussac.
7. The discovery of the chemical effects of electricity, and the decomposing agency of the Voltaic pile, by Nicholson and Carlisle; the investigation of the laws of such decompositions, by Berzelius and Hisinger: the decomposition of the alkalies by Davy, and the consequent introduction into chemistry of new and powerful agents in their metallic bases.
8. The application of chemical analysis to all the objects of organized and unorganized nature, and the discovery of the ultimate constituents of all, and the proximate ones of organic matter, and the recognisance of the important distinctions which appear to divide these great classes of bodies from each other.
9. The applications of chemistry to innumerable processes in the arts, and among other useful purposes to the discovery of the essential medical principles in vegetables, and to important medicaments in the mineral kingdom.
10. The establishment of the intimate connection

between chemical composition and crystalline form, by Haüy and Vauquelin, with the successive rectifications the statement of that connection has undergone in the hands of Mitscherlich, Rose, and others, with the progress of chemical and crystallographical knowledge.

(338.) To pursue these several heads into detail would lead us into a treatise on chemistry; but a few remarks on one or two of them, as they bear upon the general principles of all scientific enquiry, will not be irrelevant. And first, then, with reference to the discovery of new elements, it will be observed, that philosophical chemistry no more aims at determining the one essential element out of which all matter is framed — the one ultimate principle of the universe — than astronomy at discovering the origin of the planetary movements in the application of a determinate projectile force in a determinate direction, or geology at ascending to the creation of the earth. There may be such an element. Some singular relations which have been pointed out in the atomic weights of bodies seem to suggest to minds fond of speculation that there is; but philosophical chemistry is content to wait for some striking fact, which may either occur unexpectedly or be led to by the slow progress of enlarged views, to disclose to us its existence. Still, the multiplication of so-considered elementary bodies has been considered by some as an inconvenience. We confess we do not coincide with this view. Whatever they be, the obstinacy with which they resist decomposition shows that they are ingredients of a very high and primary importance in the economy of nature; and such as, in any state of science, it would be indispensably necessary to be perfectly familiar with. Like particular theorems in geometry, which, though not rising to the highest point of generality, have yet their several scopes and ranges of extensive application, they must be well and perfectly understood in all their bearings. Should we ever arrive at an analysis of these bodies, the chemical properties of the new elements which will then come into view will be known only by our knowledge of these, or of other compounds of the same class, which they may be capable of forming. Not but that such an analysis would be a most important and indeed triumphant achievement, and change the face of chemistry; but it would undo nothing that has been done, and render useless no point of knowledge which we have yet arrived at.

(339.) The atomic theory, or the law of definite proportions, which is the same thing presented in a form divested of all hypothesis, after the laws of mechanics, is, perhaps, the most important which the study of nature has yet disclosed. The extreme simplicity which characterizes it, and which is itself an indication, not unequivocal, of its elevated rank in the scale of physical truths, had the effect of causing it to be announced at once by Mr. Dalton, in its most general terms, on the contemplation of a few instances*, without passing through subordinate stages of painful inductive ascent by the intermedium of subordinate laws, such as, had the contrary course been pursued by him, would

* Thomson's First Principles of Chemistry, Introduction.

have been naturally preparatory to it, and such as would have led others to it by the prosecution of Wenzel's and Richter's researches, had they been duly attended to. This is, in fact, an example, and a most remarkable one, of the effect of that natural propensity to generalize and simplify (noticed in 171.), which, if it occasionally leads to over-hasty conclusions, limited or disproved by further experience, is yet the legitimate parent of all our most valuable and our soundest results. Instances like this, where great and. indeed, immeasurable steps in our knowledge of nature are made at once, and almost without intellectual effort, are well calculated to raise our hopes of the future progress of science, and, by pointing out the simplest and most obvious combinations as those which are actually found to be most agreeable to the harmony of creation, to hold out the cheering prospect of difficulties diminishing as we advance, instead of thickening around us in increasing complexity.

(340.) A consequence of this immediate presentation of the law of definite proportions in its most general form is, that its subordinate laws — those which limit its generality in particular cases, which diminish the number of combinations abstractly possible, and restrain the indiscriminate mixture of elements, — remain to be discovered. Some such limitations have, in fact, been traced to a certain extent, but by no means so far as the importance of the subject requires; and we have here abundant occupation for chemists for some time.

(341.) The determination of the atomic weights of the chemical elements, like that of other standard physical data, with the utmost exactness, is in itself a branch of enquiry not only of the greatest importance, but of extreme difficulty. Independent of the general reasons for desiring accuracy in this respect, there is one peculiar to the subject. It has been suggested (by Dr. Prout), and strongly insisted on (by Dr. Thomson), that all the numbers representing these weights, constituting a scale of great extent, in which the extremes already known are in proportion to each other, as 1 to upwards of 200, are simple even multiples of the least of them. If this be really the case, it opens views of such importance as to justify any degree of labour and pains in the verification of the law as a purely inductive one. But in the actual state of chemical analysis, with all deference to such high authority, we confess it appears to us to stand in great need of further confirmation, since it seems doubtful whether such accuracy has yet been attained as to enable us to answer positively for a fraction not exceeding the three or four hundredth part of the whole quantity to be determined: at least the results of the first experimenters, obtained with the greatest care, differ often by a greater amount; and this degree of exactness, at least, would be required to verify the law satisfactorily in the higher parts of the scale.

(342.) The mere agitation of such a question, however, points out a class of phenomena in physical science of a remote and singular kind, and of a very high and refined order, which could never become known but in an advanced state of science, not only practical, but theoretical, — we mean, such as consist in observed relations among the *data* of physics, which show them to be quantities not *arbitrarily* assumed, but depending on laws and causes which they may be the means of at length disclosing. A remarkable instance of such a relation is the curious law which Bode observed to obtain in the progression of the magnitudes of the several planetary orbits. This law was interrupted between Mars and Jupiter, so as to induce him to consider a planet as wanting in that interval; — a deficiency long afterwards strangely supplied by the discovery of *four* new planets in that very interval, all of whose orbits conform in dimension to the law in question, within such moderate limits of error as may be due to causes independent of those on which the law itself ultimately rests.

(343.) Neither is it irrelevant to our subject to remark, that the progress which has been made in this department of chemistry, and the considerable exactness actually attainable in chemical analysis, have been owing, in great measure, to a circumstance which might at first have been hardly considered likely to exercise much influence on the progress of a science, — the discovery of platina. Without the resources placed at the ready disposal of chemists by this invaluable metal, it is difficult to conceive that the multitude of delicate analytical experiments which have been required to construct the fabric of existing knowledge could have ever been performed. This, among many such lessons, will teach us that the most important uses of natural objects are not those which offer themselves to us most obviously. The chief use of the moon for man's immediate purposes remained unknown to him for five thousand years from his creation. And, since it cannot but be that innumerable and most important uses remain to be discovered among the materials and objects already known to us, as well as among those which the progress of science must hereafter disclose, we may hence conceive a well-grounded expectation, not only of constant increase in the physical resources of mankind, and the consequent improvement of their condition, but of continual accessions to our power of penetrating into the arcana of nature, and becoming acquainted with her highest laws.

CHAP. V.

OF THE IMPONDERABLE FORMS OF MATTER.

Heat.

(344.) One of the chief agents in chemistry, on whose proper application and management the success of a great number of its enquiries depends, and many of whose most important laws are disclosed to us by phenomena of a chemical nature, is HEAT. Although some of its effects are continually before our eyes as matters of the most common occurrence, insomuch that there is scarcely any process in the useful arts and manufactures which does not call for its intervention, and although, independent of this high utility, and the proportionate importance of a knowledge of its nature and laws, it presents in itself a subject of the most curious speculation; yet there is scarcely any physical agent of which we have so imperfect a knowledge, whose intimate nature is more hidden, or whose laws are of such delicate and difficult investigation.

(345.) The word heat generally implies the sensation which we experience on approaching a fire; but, in the sense it carries in physics, it denotes the cause, whatever it be, of that sensation, and of all the other phenomena which arise on the application of fire, or of any other heating cause. We should be greatly deceived if we referred only to sensation as an indication of the presence of this cause. Many of those things which excite in our organs, and especially of those of taste, a sensation of heat, owe this property to chemical stimulants, and not at all to their being actually *hot*. This error of judgment has produced a corresponding confusion of language, and hence had actually at one period * crept into physical philosophy a great many illogical and absurd conclusions. Again, there are a number of chemical agents, which, from their corroding, blackening, and dissolving, or drying up the parts of some descriptions of bodies, and producing on them effects not generally unlike (though intrinsically very different from) those produced by heat, are said, in loose and vulgar language, to burn them; and this error has even become rooted into a prejudice, by the fact that some of these agents are capable of becoming actually and truly *hot* during their action on moist substances, by reason of their combination with the water the latter contain. Thus, quicklime and oil of vitriol both exercise a powerful corrosive action on animal and vegetable substances, and both become violently hot by their combination with water. They are, therefore, set down in vulgar parlance as substances of a hot nature; whereas, in their relations to the physical cause of heat, they agree with the generality of bodies similarly constituted.

* Novum Organum, part ii. table 2. (24), (30), &c. on the form or nature of heat.

(346.) The nature of heat has hitherto been chiefly studied under the general heads of—

1st, Its sources, or the phenomena which it usually accompanies.

2d, Its communication from its sources to substances capable of receiving it, and from these to others, with a view to discover the laws which regulate its distribution through space or through the bodies which occupy it.

3d, Its effects, on our senses, and on the bodies to which it is communicated in its various degrees of intensity, by which, means are afforded us of measuring these degrees.

4th, Its intimate relations to the atoms of matter, as exhibited in its capability of acquiring a latent state under certain circumstances, and of entering into something like chemical combinations.

(347.) The most obvious sources of heat are, the sun, fire, animal life, fermentations, violent chemical actions of all kinds, friction, percussion, lightning, or the electric discharge, in whatever manner produced, the sudden condensation of air, and others, so numerous, and so varied, as to show the extensive and important part it has to perform in the economy of nature. The discoveries of chemists, however, have referred most of these to the general head of chemical combination. Thus, fire, or the combustion of inflammable bodies, is nothing more than a violent chemical action attending the combination of their ingredients with the oxygen of the air. Animal heat is, in like manner, referable to a process bearing no remote analogy to a slow combustion, by which a portion of *carbon*, an inflammable principle existing in the blood, is united with the oxygen of the air in respiration; and thus carried off from the system: fermentation is nothing more than a decomposition of chemical elements loosely united, and their re-union in a more permanent state of combination. The analogy between the sun and terrestrial fire is so natural as to have been chosen by Newton to exemplify the irresistible force of an inference derived from that principle. But the nature of the sun and the mode in which its wonderful supply of light and heat is maintained are involved in a mystery which every discovery that has been made either in chemistry or optics, so far from elucidating, seems only to render more profound. Friction as a source of heat is well known: we rub our hands to warm them, and we grease the axles of carriage-wheels to prevent their setting fire to the wood; an accident which, in spite of this precaution, does sometimes happen. But the effect of friction, as a means of producing heat with little or no consumption of materials, was not fully understood till made the subject of direct experiment by count Rumford, whose results appear to have established the extraordinary fact, that an unlimited supply of heat may be derived by friction from the same materials. Condensation, whether of air by pressure, or of metals by percussion, is another powerful source of heat. Thus, iron may be so dexterously hammered as to become red-hot, and the rapid condensation of a confined portion of air will set tinder on fire.

(348.) The most violent heats known are produced by the concentration of the solar rays by burning glasses,—by the combustion of oxygen and hydrogen gases mixed in the exact proportion in which they combine to produce water,—and by the discharge of a continued and copious current of electricity through a small conductor. As these three sources of heat are independent of each other, and each capable of being brought into action in a very confined space, there seems no reason why they might not all three be applied at once at the same point, by which means, probably, effects would be produced infinitely surpassing any hitherto witnessed.

(349.) Heat is communicated either by *radiation* between bodies at a distance, or by *conduction* between bodies in contact, or between the contiguous parts of one and the same body. The laws of the radiation of heat have been studied with great attention, and have been found to present strong analogies with that of light in some points, and singular differences in others. Thus, the heat which accompanies the sun's rays comports itself, in all respects, like light; being subject to similar laws of reflection, refraction, and even of polarization, as has been shown by Berard. Yet they are not identical with each other; Sir William Herschel having shown, by decisive experiments, verified by those of Sir H. Englefield, that there exist in a solar beam both rays of heat which are not luminous, and rays of light which have no heating power.

(350.) The heat, radiated by terrestrial fires, and by bodies *obscurely* hot, by whatever means they have acquired their heat (even by exposure to the sun's rays), differs very materially from solar heat in their power of penetrating transparent substances. This singular and important difference was first noticed by Mariotte, and afterwards made the subject of many curious and interesting experiments by Scheele, who found that terrestrial heat, or that radiated from fires or heated bodies, is intercepted and detained by glass or other transparent bodies, while solar heat is not; and that, being so detained, it heats them: which the latter, as it passes freely through them, is incapable of doing. The more recent researches of Delaroche, however, have shown that this detention is complete only when the temperature of the source of heat is low; but that, as that temperature is higher, a portion of the heat radiated acquires a power of penetrating glass; and that the quantity which does so bears continually a larger and larger proportion to the whole, as the heat of the radiant body is more intense. This discovery is very important, as it establishes a community of nature between solar and terrestrial heat; while at the same time it leads us to regard the actual temperature of the sun as far exceeding that of any earthly flame.

(351.) A variety of theories have been framed to account for these curious phenomena; but the subject stands rather in need of further elucidation from experiment, and is one which merits, and will probably amply repay, the labours of those who may hereafter devote their attention to it. The theory f the radiation of heat, in general, which seems to agree best with the known phenomena, is that of M. Prevost, who considers all bodies as constantly radiating out heat in all directions, and receiving it by a similar means of communication from others, and thus tending, in any space filled, wholly or in part, with bodies at various temperatures, to establish an equilibrium or equality of heat in all parts. The application of this idea to the explanation of the phenomenon of dew we have already seen (see 167.). The laws of such radiation, under various circumstances, have been lately investigated in a beautiful series of experiments on the cooling of bodies by their own radiation in vacuo, by Messrs. Dulong and Petit, which offer some of the best examples in science of the inductive investigation of quantitative laws.

(352.) The communication of heat between bodies in contact, or between the different parts of the same body, is performed by a process called conduction. It is, in fact, only a particular case of radiation, as has been explained above (217.); but a case *so* particular as to require a separate and independent investigation of its laws. The most important consideration introduced into the enquiry by this peculiarity is that of time. The communication of heat by conduction is performed, for the most part, with extreme slowness, while that performed by direct radiation is probably not less rapid than the propagation of light itself. The analysis of the delicate and difficult points which arise in the investigation of this subject in its reduction to direct geometrical treatment has been executed with admirable success by the late Baron Fourrier, whose recent lamented death has deprived science of an ornament it could ill spare, thinned as its ranks have been within the last few years. This acute philosopher and profound mathematician has developed, in a series of elaborate memoirs presented to the French Institute, the laws of the communication of heat through the interior of solid masses, placed under the influence of any external heating and cooling causes, and has in particular applied his results to the conditions on which the maintenance of the actual observed temperature on the earth's surface depends; to the possible influence of a supposed central heat on our climates; and to the determination of the actual amount of the heat, derived to us from the sun, or at least that portion of it on which the difference of the seasons depends.

(353.) The principal effects of heat are the sensations of warmth or cold consequent on its entry or egress into or out of our bodies; the dilatation it causes in the dimensions of all substances in which it is accumulated; the changes of state it produces in the melting of solids, and the conversion of them and of liquids into vapour; and the chemical changes it performs by actual decompositions effected in the intimate molecules of various substances, especially those of which vegetables and animals are composed; to which we may add, the production of electric phenomena under certain circumstances in the contact of metals, and the developement of electric polarity in crystallised substances.

(354.) Cold has been considered by some as a positive quality, the effect of a cause antagonist to that of heat; but this idea seems now (with perhaps a single exception) to be universally abandoned. The sensation of cold is as easily explicable by the passage of heat outwards through the surface of the body as that of heat by its ingress from without; and the experiments cited in proof of a radiation of cold are all perfectly explained by Prevost's theory of reciprocal interchange. It is remarkable, however, how very limited our means of producing intense cold are, compared with those we possess of effecting the accumulation of heat in bodies. This is one of the strongest arguments adducible in favour of the doctrines of those who maintain the possibility of exhausting the heat of a body altogether, and leaving it in a state absolutely devoid of it. But we ought to consider, that the known methods of generating heat chiefly turn on the production of chemical combinations: we may easily conceive, therefore, that, to obtain equally powerful corresponding frigorific effects, we ought to possess the means of effecting a disunion equally extensive and rapid between such elements, actually combined, as have already produced heat by their union. This, however, we can only accomplish by engaging them in combinations still more energetic, that is to say, in which we may reasonably expect more heat to be produced by the new combination than would be destroyed or abstracted by the proposed decomposition. Chemistry, however, (unaided by electric agency,) affords no means of suddenly breaking the union of two elements, and presenting *both* in an uncombined state. A certain analogy to such disunion, however, and its consequences, may be traced in the sudden expansion of condensed gases from a liquid state into vapour, which is the most powerful source of cold known.

(355.) The dilatation of bodies by heat forms the subject of that branch of science called pyrometry. There is no body but is capable of being penetrated by heat, though some with greater, others with less rapidity; and being so penetrated, all bodies (with a very few exceptions, and those depending on very peculiar circumstances,) are dilated by it in bulk, though with a great diversity in the amount of dilatation produced by the same degree of heat. Of the several forms of natural bodies, gases and vapours are observed to be most dilatable; liquids next, and solids least of all. The dilatation of solids has been made a subject of repeated and careful measurement by several experimenters; among whom, Smeaton, Lavoisier, and Laplace, are the principal. The remarkable discovery of the unequal dilatation of crystallised bodies by Mitscherlich has already been spoken of. (266.) That of gases and vapours was examined about the same time by Dalton and Gay-Lussac, who both arrived independently at the conclusion of an equal dilatability subsisting in them all, which constitutes one of the most remarkable points in their history.

(356.) The dilatation of air by heat affords, perhaps, the most unexceptionable means known of measuring degrees of heat. The thermometer, as originally constructed by Cornelius Drebell, was an air thermometer. Those now in common use measure accessions of heat not by the degree of dilatation of air but of mercury. It has been shown, by the researches of Dulong and Petit, that its indications coincide exactly with that of the air-thermometer in moderate temperatures; though at very elevated ones they exhibit a sensible, and even considerable, deviation. By this instrument, which owes its present convenience and utility to the happy idea of Newton, who first thought of fixing determinate points on its scale, we are enabled to estimate, or at least identify, the degrees of heat; and thereby to investigate with accuracy the laws of its communication and its other properties. Were we sure that equal additions of heat produced equal increments of dimension in any substance, the indications of a thermometer would afford a true and secure *measure* of the quantity present; but this is so far from being the case, that we are nearly in total ignorance on this important point; a circumstance which throws the greatest difficulty in the way of all theoretical reasoning, and even of experimental enquiry. The laws of the dilatation of liquids, in consequence of this deficiency of necessary preliminary knowledge, are still involved in great obscurity, notwithstanding the pains which have been bestowed on them by the elaborate experiments and calculations of Gilpin, Blagden, Deluc, Dalton, Gay-Lussac, and Biot.

(357.) The most striking and important of the effects of heat consist, however, in the liquefaction of solid substances, and the conversion of the liquids so produced into vapour. There is no solid substance known which, by a sufficiently intense heat, may not be melted, and finally dissipated in vapour; and this analogy is so extensive and cogent, that we cannot but suppose that all those bodies which are liquid under ordinary circumstances, owe their liquidity to heat, and would freeze or become solid if their heat could be sufficiently reduced. In many we see this to be the case in ordinary winters; for some, severe frosts are requisite; others freeze only with the most intense artificial colds; and some have hitherto resisted all our endeavours; yet the number of these last is few, and they will probably cease to be exceptions as our means of producing cold become enlarged.

(358.) A similar analogy leads us to conclude that all aëriform fluids are merely liquids kept in the state of vapour by heat. Many of them have been actually condensed into the liquid state by cold accompanied with violent pressure; and as our means of applying these causes of condensation have improved, more and more refractory ones have successively yielded. Hence we are fairly entitled to extend our conclusion to those which we have not yet been able to succeed with; and thus we are led to regard it as a general fact, that the liquid and aëriform or vaporous states are entirely dependent on *heat;* that were it not for this cause, there would be nothing but solids in nature; and that, on the other hand, nothing but a sufficient intensity of heat is requisite to destroy the cohesion of every substance, and reduce all bodies, first to liquids, and then into vapour.

(359.) But solids, themselves, by the abstraction of heat shrink in dimension, and at the same time become harder, and more brittle; yielding less to pressure, and permitting less separation between their parts by tension. These facts, coupled with the greater compressibility of liquids, and the still greater of gases, strongly induce us to believe that it is heat, and heat alone, which holds the particles of all bodies at that distance from each other which is necessary to allow of compression; which in fact gives them their elasticity, and acts as the antagonist force to their mutual attraction, which would otherwise draw them into actual contact, and retain them in a state of absolute immobility and impenetrability. Thus we learn to regard heat as one of the great maintaining powers of the universe, and to attach to all its laws and relations a degree of importance which may justly entitle them to the most assiduous enquiry.

(360.) It was first ascertained by Dr. Black that when heat produces the liquefaction of a solid, or the conversion of a liquid into vapour, the liquid or the vapour resulting is no *hotter* than the solid or liquid from which it was produced, though a great deal of heat has been expended in producing this effect, and has actually entered into the substance.

(361.) Hence he drew the conclusion that it has become *latent*, and continues to exist in the product, maintaining it in its new state, without increasing its temperature. He further proved, that when the vapour condenses, or the liquid freezes, this latent heat is again given out from it. This great discovery, with its natural and hardly less important concomitant, that of the difference of specific heats in different bodies, or the different quantities of heat they require to raise their temperature equally, are the chief reasons for regarding heat as a material substance in a more decided manner than light, with which in its radiant state it holds so close an analogy.

(362.) The subject of latent heat has been far less attentively studied than its great practical importance would appear to demand, when we consider that it is to this part of physical science that the theory of the steam-engine is mainly referable, and that material improvements may not unreasonably be expected in that wonderful instrument, from a more extended knowledge than we possess of the latent heats of different vapours. This is not the case, however, with the subject of specific heat, which was followed up immediately after its first promulgation with diligence by Irvine; and, after a brief interval, by Lavoisier and Laplace, as well as by our countryman Crawfurd, who determined the specific heats of many substances, both solid and liquid. After a considerable period of inactivity, the subject was again resumed by Delaroche and Berard, and subsequently by Dulong and Petit: the result of whose investigations has been the inductive establishment of one of those simple and elegant physical laws which carry with them, if not their own evidence, at least their own recommendation to our belief, as being in unison with every thing we know of the harmony of nature. The law to which we allude is this:— that the atoms of all the simple chemical elements have exactly the same capacity for heat, or are all equally heated or cooled by equal accessions or abstractions of heat. It is only among laws like this that we can expect to find a clew capable of guiding us to a knowledge of the true nature of heat, and its relations to ponderable matter.

Magnetism and Electricity.

(363.) These two subjects, which had long maintained a distinct existence, and been studied as separate branches of science, are at length effectually blended. This is, perhaps, the most satisfactory result which the experimental sciences have ever yet attained. All the phenomena of magnetic polarity, attraction, and repulsion, have at length been resolved into one general fact, that two currents of electricity, moving in the same direction repel, and in contrary directions attract, each other. The phenomena of the communication of magnetism and what is called its induced state, alone remain unaccounted for; but the interesting theory which has been developed by M. Ampere, under the name of Electro-dynamics, holds out a hope that this difficulty will also in its turn give way, and the whole subject be at length completely merged, as far as the consideration of the acting causes goes, in the more general one of electricity. This, however, does not prevent magnetism from maintaining its separate importance as a department of physical enquiry, having its own peculiar laws and relations of the highest practical interest, which are capable of being studied quite apart from all consideration of its electrical origin. And not only so, but to study them with advantage, we must proceed as if that origin were totally unknown, and, at least up to a certain point, and that a considerably advanced one, conduct our enquiries into the subject on the same inductive principles as if this branch of physics were absolutely independent of all others.

(364.) Iron, and its oxides and alloys, were for a long time the only substances considered susceptible of magnetism. The loadstone was even one of the examples produced by Bacon of that class of physical instances to which he applies the term "Instantiæ monodicæ" — *singular instances.* And the history of magnetism affords a beautiful comment on his remark on instances of this sort. "Nor should our enquiries," he observes, "into their nature be broken off, till the properties and qualities found in such things as may be esteemed wonders in nature are reduced and comprehended under some certain law; so that all irregularity or singularity may be found to depend upon some common form, and the wonder only rest in the exact differences, degrees, or extraordinary concurrence, and not in the species itself." The discovery of the magnetism of nickel, which though inferior to that of iron, is still considerable; that of cobalt, yet feebler, and that of titanium, which is only barely perceptible, have effectually broken down the imaginary limit between iron and the other materials of the world, and established the existence of that general law of continuity which it is one chief business of philosophy to trace throughout nature. The more recent discoveries of M. Arago (mentioned in 160.) have completed this generalization, by showing that there is no substance but which, under proper circumstances, is capable of exhibiting unequivocal signs of the magnetic virtue. And to obliterate all traces of that line of separation which was once so broad, we are now enabled, by the great discovery of Oersted, to communicate at and during pleasure to a coiled wire of any metal indifferently all the properties of a magnet;—its attraction, repulsion, and polarity; and *that* even in a more intense degree than was previously thought to be possible in the best natural magnets. In short, in this case, and in this case only, perhaps, in science, have we arrived at that point which Bacon seems to have understood by the discovery of "forms." "The *form* of any nature," says he, "is such, that where *it* is, the given nature must infallibly be. The form, therefore, is perpetually present when that nature is present; ascertains it universally, and accompanies it every where. Again, this form is such, that when removed, the given nature infallibly vanishes. Lastly, a true form is such as can deduce a given nature from some essential property, which resides in many things."

(365.) Magnetism is remarkable in another important point of view. It offers a prominent, or "*glaring instance*" of that quality in nature which is termed *polarity* (267.), and that under circumstances which peculiarly adapt it for the study of this quality. It does not appear that the ancients had any knowledge of this property of the magnet, though its attraction of iron was well known to them. The first mention of it in modern times cannot be traced earlier than 1180, though it was probably known to the Chinese before that time. The polarity of the magnet consists in this, that if suspended freely, one part of it will invariably direct itself towards a certain point in the horizon, the other towards the opposite point; and that, if two magnets, so suspended, be brought near each other, there will take place a mutual action, in consequence of which, the positions of both will be disturbed, in the same manner as would happen if the corresponding parts of each repelled, and those oppositely directed attracted, each other; and by properly varying the experiment, it is found that they really do so. If a small magnet, freely suspended, be brought into the neighbourhood of a larger one, it will take a position depending on the position of the *poles* of the larger one, with respect to its point of suspension. And it has been ascertained that these and all other phenomena exhibited by magnets in their mutual attractions and repulsions are explicable on the supposition of two forces or virtues lodged in the particles of the magnets, the one predominating at one end, the other at the other; and such that each particle shall attract those in which the *opposite* virtue to its own prevails, and repel those in which a *similar* one resides with a force proportional to the inverse square of their mutual distance.

(366.) The direction in which a magnetic bar, or needle of steel, freely suspended, places itself, has been ascertained to be different at different points of the earth's surface. In some places it points exactly north and south, in others it deviates from this direction more or less, and at some actually stands at right angles to it. This remarkable phenomenon, which is called the variation of the needle, and which was discovered by Sebastian Cabot in the year 1500, is accompanied with another called the dip, noticed by Robert Norman in 1576. It consists in a tendency of a needle, nicely balanced on its centre, when unmagnetized, to *dip* or point downwards when rendered magnetic, towards a point below the horizon, and situated within the earth. By tracing the variation and dip over the whole surface of the globe, it has been found that these phenomena take place as they would do if the earth itself were a great magnet, having its poles deeply situated below the surface,—and, what is very remarkable, possessing a slow motion within it, in consequence of which neither the variation nor dip remain constantly the same at the same place. The laws of this motion are at present unknown; but the discovery of electro-magnetism, by rendering it almost certain that the earth's magnetism is merely an effect of the continual circulation of great quantities of electricity round it, in a direction generally corresponding with that of its rotation, have dissipated the greater part of the mystery which hung over these phenomena; since a variety of causes, both geological and others, may be imagined which may produce considerable deviations in the intensity, and partial ones in the direction, of such electric currents. The unequal distribution of land and sea in the two hemispheres, by affecting the operation of the sun's heat in producing evaporation from the latter, which is probably one of the great sources of terrestrial electricity, may easily be conceived to modify the general tendency of such currents, and to produce irregularities in them, which may render a satisfactory account of whatever still appears anomalous in the phenomena of terrestrial magnetism. This branch of science thus becomes connected, on a great scale, with that of meteorology, one of the most complicated and difficult, but at the same time interesting, subjects of physical research; one, however, which has of late begun to be studied with a diligence which promises the speedy disclosure of relations and laws of which at present we can form but a very imperfect notion.

(367.) The communication of magnetism from the earth to a magnetic body, or from one magnetic body to another, is performed by a process to which the name of induction has been given, and the laws and properties of such induced magnetism have been studied with much perseverance and success,—practically, by Gilbert, Boyle, Knight, Whiston, Cavallo, Canton, Duhamel, Rittenhouse, Scoresby, and others; and theoretically, by Æpinus, Coulomb, and Poisson, and in our own country by Messrs. Barlow and Christie, who have investigated with great care the curious phenomena which take place when masses of iron are presented successively, in different positions, by rotation on an axis, to the influence of the earth's magnetism. The magnetism of crystallized bodies (partly from the extreme rarity of such as are susceptible of any considerable magnetic virtue) has not hitherto been at all examined, but would probably afford very curious results.

(368.) To electricity the views of the physical enquirer now turn from almost every quarter, as to one of those universal powers which Nature seems to employ in her most important and secret operations. This wonderful agent, which we see in intense activity in lightning, and in a feebler and more diffused form traversing the upper regions of the atmosphere in the northern lights, is present, probably in immense abundance, in every form of matter which surrounds us, but becomes sensible only when disturbed by excitements of peculiar kinds. The most effectual of these is friction, which we have already observed to be a powerful source of heat. Everybody is familiar with the crackling sparks which fly from a cat's back when stroked. These, by proper management, may be accumulated in bodies suitably disposed to receive them, and, although then no longer visible, give evidence of their existence by the exhibition of a vast variety of extraordinary phenomena, — producing attractions and repulsions in bodies at a distance, — admitting of being transferred by contact, or by sudden and violent transilience of the interval of separation, from one body to another, under the form of sparks and flashes ; — traversing with perfect facility the substance of the densest metals, and a variety of other bodies called conductors, but being detained by others, such as glass, and especially *air*, which are thence called non-conductors, — producing painful shocks and convulsive motions, and even death itself if in sufficient quantity, in animals through which they pass, and finally imitating, on a small scale, all the effects of lightning.

(369.) The study of these phenomena and their laws until a comparatively recent period occupied the entire attention of electricians, and constituted the whole of the science of electricity. It appears, as the result of their enquiries, that all the phenomena in question are explicable on the supposition that electricity consists in a rare, subtle, and highly elastic fluid, which in its tendency to expand and diffuse itself pervades with more or less facility the substance of conductors, but is obstructed and detained from expansion more or less completely by non-conductors. It is supposed, moreover, that this electric fluid possesses a power of attraction for the particles of all ponderable matter, together with that of a repulsion for particles of its own kind. Whether it has weight, or is rather to be regarded as a species of matter distinct from that of which ponderable bodies consist, is a question of such delicacy, that no direct experiments have yet enabled us to decide it; but at all events its *inertia* compared with its elastic force must be conceived excessively small, so that it is to be regarded as a fluid in the highest degree *active*, obeying every impulse, internal or external, with the greatest promptitude; in short, a fluid whose energies can only be compared with those of the ethereal medium by which, in the undulatory doctrine, light is supposed to be conveyed. The properties of hydrogen gas compared with those of the denser aëriform fluids will, in some slight degree, aid our conception of the excessive mobility and penetrating activity of a fluid so constituted. Electricity, however, must be regarded as differing in some remarkable points from all those fluids to which we have hitherto been accustomed to apply the epithet elastic, such as air, gases, and vapours. In these, the repulsive force of the particles on which their elasticity depends is considered as extending only to very small distances, so as to affect only those in the immediate vicinity of each other, while their attractive power, by which they obey the general gravitation of all matter, extends to any distance. In electricity, on the other hand, the very reverse must be admitted.

The force by which its particles repel each other extends to great distances, while its force of adhesion to ponderable matter must be regarded as limited in its extent to such minute intervals as escape observation.

(370.) The conception of a single fluid of this kind, which when accumulated in excess in bodies tends constantly to escape, and seek a restoration of equilibrium by communicating itself to any others where there may be a deficiency, is that which occurs most naturally to the mind, and was accordingly maintained by Franklin, to whom the science of electricity is under great obligations for those decisive experiments which informed us respecting the true nature of lightning. The same theory was afterwards advocated by Æpinus, who first showed how the laws of equilibrium of such a fluid might be reduced to strict mathematical investigation. But there are phenomena accompanying its transfer from body to body and the state of equilibrium it affects under various circumstances, which appear to require the admission of *two distinct fluids* antagonist to each other, each attracting the other, and repelling itself; but each, alike, susceptible of adhesion to material substances, and of transfer more or less rapid from particle to particle of them. These fluids in the natural undisturbed state are conceived to exist in a state of combination and mutual saturation; but this combination may be broken, and either of them separately accumulated in a body to any amount without the other, provided its escape be properly obstructed by surrounding it with non-conductors. When so accumulated, its repulsion for its own kind and attraction of the opposite species in neighbouring bodies tends to disturb the natural equilibrium of the two fluids present in them, and to produce phenomena of a peculiar description, which are termed *induced* electricity. Curious and artificial as this theory may appear, there has hitherto been produced no phenomenon of which it will not afford at least a plausible, and in by far the majority of cases a very satisfactory, explanation. It has one character which is extremely valuable in any theory, that of admitting the application of strict mathematical reasoning to the conclusions we would draw from it. Without this, indeed, it is scarcely possible that any theory should ever be fairly brought to the test by a comparison with facts. Accordingly, the mathematical theory of electrical equilibrium, and the laws of the distribution of the electric fluids over the surfaces of bodies in which they are accumulated, have been made the subject of elaborate geometrical investigation by the most expert mathematicians, and have attained a degree of extent and elegance which places this branch of science in a very high rank in the scale of mathematico-physical enquiry. These researches are grounded on the assumption of a law of attraction and repulsion similar to those of gravity and magnetism, and which by the general accordance of the results with facts, as well as by experiments instituted for the express purpose of ascertaining the laws in question, are regarded as sufficiently demonstrated.

(371.) The most obscure part of the subject is no

doubt the original mode of disturbance of electrical equilibrium, by which electricity is excited in the first instance, either by friction or by any other of those causes which have been ascertained to produce such an effect: analogies, it is true, are not wanting*; but it must be allowed that hitherto

* We will mention one which we do not remember to have seen noticed elsewhere in the case of a disturbance of the equilibrium of heat produced by means purely mechanical, and by a process depending entirely on a certain order and sequence of events, and the operation of known causes. Suppose a quantity of air enclosed in a metallic reservoir, of some good conductor of heat, and suddenly compressed by a piston. After giving time for the heat developed by the condensation to be communicated from the air to the metal which will be thereby more or less raised in temperature *above* the surrounding atmosphere, let the piston be suddenly retracted and the air restored to its original volume in an instant. The whole apparatus is now precisely in its initial situation, as to the disposition of its material parts, and the whole quantity of heat it contains remains unchanged. But it is evident that the distribution of this heat within it is now very different from what it was before; for the air in its sudden expansion cannot re-absorb in an instant of time all the heat it had parted with to the metal: it will, therefore, have a temperature *below* that of the general atmosphere, while the metal yet retains one above it. Thus, a subversion of the equilibrium of temperature has been *bonâ fide* effected. Heat has been driven from the air into the metal, while every thing else remains unchanged.

We have here a means by which, it is evident, heat may be obtained, to any extent, from the air, without fuel. For if, in place of withdrawing the piston and letting the *same* air expand, within the reservoir, it be allowed to escape so suddenly as not to re-absorb the heat given off, and fresh air be then admitted and the process repeated, any quantity of air may thus be *drained* of its heat.

nothing decisive has been offered on the subject; and that conjectural modes of action have in this instance too often usurped the place of those to which a careful examination of facts alone can lead us.

(372.) Philosophers had long been familiar with the effects of electricity above referred to, and with those which it produces in its sudden and violent transfer from one body to another, in rending and shattering the parts of the substances through which it passes, and where in great quantity, producing all the effect of intense heat, igniting, fusing, and volatilizing metals, and setting fire to inflammable bodies; even its occasional influence in destroying or altering the polarity of the magnetic needle had been noticed: but as heat was known to be produced by mechanical violence, and as magnetism was also known to be greatly affected by the same cause, these effects were referred rather to that cause than to any thing in the peculiar nature of the electric matter, and regarded rather as an indirect consequence of its mode of action than as connected with its intimate nature. In short, electricity seemed destined to furnish another in addition to many instances of subjects insulated from the rest of philosophy, and capable of being studied only in its own internal relations, when the great discoveries of Galvani and Volta placed a new power at the command of the experimenter, by whose means those effects which had before been crowded within an inappreciable instant could be developed in detail and studied at leisure; and those forces which had previously exhibited themselves only in a state of uncontrollable intensity were tamed down, as it were, and made to distribute their efficacy over an indefinite time, and to regulate their action at the will of the operator. It was then soon ascertained that electricity in the act of its passage along conductors, produces a variety of wonderful effects, which had never been previously suspected; and these of such a nature, as to afford points of contact with several other branches of physical enquiry, and to throw new and unexpected lights on some of the most obscure operations of nature.

(373.) The history of this grand discovery affords a fine illustration of the advantage to be derived in physical enquiry from a close and careful attention to any phenomenon, however apparently trifling, which may at the moment of observation appear inexplicable on received principles. The convulsive motions of a dead frog in the neighbourhood of an electric discharge, which originally drew Galvani's attention to the subject, had been noticed by others nearly a century before his time, but attracted no further remark than as indicating a peculiar sensibility to electrical excitement depending on that remnant of vitality which is not extinguished in the organic frame of an animal by the deprivation of actual life. Galvani was not so satisfied. He analysed the phenomenon; and in investigating all the circumstances connected with it was led to the observation of a peculiar electrical excitement which took place when a circuit was formed of three distinct parts, a muscle, a nerve, and a metallic conductor, each placed in contact with the other two, and which was manifested by a convulsive motion produced in the muscle. To this phenomenon he gave the name of animal electricity, an unfortunate epithet, since it tended to restrict enquiry into its nature to the class of phenomena in which it first became apparent. But this circumstance, which in a less enquiring age of science might have exercised a fatal influence on the progress of knowledge, proved happily no obstacle to the further developement of its principles, the subject being immediately taken up with a kind of prophetic ardour by Volta, who at once generalized the phenomena, rejecting the physiological considerations introduced by Galvani, as foreign to the enquiry, and regarding the contraction of the muscles as merely a delicate means of detecting the production of electrical excitements too feeble to be rendered sensible by any other means. It was thus that he arrived at the knowledge of a general fact, that of the disturbance of electrical equilibrium by the mere contact of different bodies, and the circulation of a current of electricity in one constant direction, through a circuit composed of three different conductors. To increase the intensity of the very minute and delicate effect thus observed became his next aim, nor did his enquiry terminate till it had placed him in possession of that most wonderful of all human inventions, the pile which bears his name, through the medium of a series of well conducted and logically combined experiments, which has rarely, if ever, been surpassed in the annals of physical research.

(374.) Though the original pile of Volta was feeble

compared to those gigantic combinations which were afterwards produced, it sufficed, however, to exhibit electricity under a very different aspect from any thing which had gone before, and to bring into view those peculiar modifications in its action which Dr. Wollaston was the first to render a satisfactory account of, by referring them to an increase of *quantity*, accompanied with a diminution of *intensity* in the supply afforded. The discovery had not long been made public, and the instrument in the hands of chemists and electricians, before it was ascertained that the electric current, transmitted by it through conducting liquids, produces in them chemical decompositions. This capital discovery appears to have been made, in the first instance, by Messrs. Nicholson and Carlisle, who observed the decomposition of water so produced. It was speedily followed up by the still more important one of Berzelius and Hisinger, who ascertained it as a general law, that, in all the decompositions so effected, the acids and oxygen become transferred to, and accumulated around, the positive, — and hydrogen, metals, and alkalies round the negative, pole of a Voltaic circuit; being transferred in an invisible, and, as it were, a latent or torpid state, by the action of the electric current, through considerable spaces, and even through large quantities of water or other liquids, again to re-appear with all their properties at their appropriate resting-places.

(375.) It was in this state of things that the subject was taken up by Davy, who, seeing that the strongest chemical affinities were thus readily subverted by the decomposing action of the pile, conceived the happy idea of bringing to bear the intense power of the enormous batteries of the Royal Institution on those substances which, though strongly suspected to be compounds, had resisted all attempts to decompose them — the alkalies and earths. They yielded to the force applied, and a total revolution was thus effected in chemistry; not so much by the introduction of the new elements thus brought to light, as by the mode of conceiving the nature of chemical affinity, which from that time has been regarded (as Davy broadly laid it down, in a theory which was readily adopted by the most eminent chemists, and by none more readily than by Berzelius himself,) as entirely due to electric attractions and repulsions, those bodies combining most intimately whose particles are habitually in a state of the most powerful electrical antagonism, and dispossessing each other, according to the amount of their difference in this respect.

(376.) The connection of magnetism and electricity had long been suspected, and innumerable fruitless trials had been made to determine, in the affirmative or negative, the question of such connection. The phenomena of many crystallized minerals which become electric by heat, and develope opposite electric poles at their two extremities, offered an analogy so striking to the polarity of the magnet, that it seemed hardly possible to doubt a closer connection of the two powers. The developement of a similar polarity in the Voltaic pile pointed strongly to the same conclusion; and experiments had even been made with a view to ascertain whether a pile in a state of excitement might not manifest a disposition to place itself in the magnetic meridian; but the essential condition had been omitted, that of allowing the pile to discharge itself freely, a condition which assuredly never would have occurred of itself to any experimenter. Of all the philosophers who had speculated on this subject, none had so pertinaciously adhered to the idea of a necessary connection between the phenomena as Oersted. Baffled often, he returned to the attack; and his perseverance was at length rewarded by the complete disclosure of the wonderful phenomena of electro-magnetism. There is something in this which reminds us of the obstinate adherence of Columbus to his notion of the necessary existence of the New World; and the whole history of this beautiful discovery may serve to teach us reliance on those general analogies and parallels between great branches of science by which one strongly reminds us of another, though no direct connection appears; as an indication not to be neglected of a community of origin.

(377.) It is highly probable that we are still ignorant of many interesting features in electrical science, which the study of the Voltaic circuit will one day disclose. The violent mechanical effects produced by it on mercury, placed under conducting liquids which have been referred by Professor Erman to a modified form of capillary attraction, but which a careful and extended view of the phenomena have led others* to regard in a very different light, as pointing out a primary action of a dynamical rather than a statical character, deserve, in this point of view, a further investigation; and

* See Phil. Trans. 1824.

the curious relations of electricity to heat, as exhibited in the phenomena of what has been called thermo-electricity, promise an ample supply of new information.

(378.) Among the remarkable effects of electricity disclosed by the researches of Galvani and Volta, perhaps the most so consisted in its influence on the nervous system of animals. The origin of muscular motion is one of those profound mysteries of nature which we can scarcely venture to hope will ever be fully explained. Physiologists, however, had long entertained a general conception of the conveyance of some subtle fluid or spirit from the brain to the muscles of animals along the nerves; and the discovery of the rapid transmission of electricity along conductors, with the violent effects produced by shocks, transmitted through the body, on the nervous system, would very naturally lead to the idea that this nervous fluid, if it had any real existence, might be no other than the electrical. But until the discoveries of Galvani and Volta, this could only be looked upon as a vague conjecture. The character of a *vera causa* was wanting to give it any degree of rational plausibility, since no reason could be imagined for the disturbance of the electrical equilibrium in the animal frame, composed as it is entirely of conductors, or rather, it seemed contrary to the then known laws of electrical communication to suppose any such. Yet one strange and surprising phenomenon might be adduced indicative of the possibility of such disturbance, viz. the powerful shock given by the torpedo and other fishes of the same kind, which presented so many analogies with those arising from electricity, that they could hardly be referred to a different source, though *besides* the shock neither spark nor any other indication of electrical tension could be detected in them.

(379.) The benumbing effect of the torpedo had been ascertained to depend on certain singularly constructed organs composed of membranous columns, filled from end to end with laminæ, separated from each other by a fluid: but of its mode of action no satisfactory account could be given; nor was there any thing in its construction, and still less in the nature of its materials, to give the least ground for supposing it an electrical apparatus. But the pile of Volta supplied at once the analogies both of structure and of effect, so as to leave little doubt of the electrical nature of the apparatus, or of the power, a most wonderful one certainly, of the animal, to determine, by an effort of its will, that concurrence of conditions on which its activity depends. This remained, as it probably ever will remain, mysterious and inexplicable; but the principle once established, that there exists in the animal economy a power of determining the developement of electric excitement, capable of being transmitted along the nerves, and it being ascertained, by numerous and decisive experiments, that the transmission of Voltaic electricity along the nerves of even a dead animal is sufficient to produce the most violent muscular action, it became an easy step to refer the origin of muscular motion in the living frame to a similar cause; and to look to the brain, a wonderfully constituted organ, for which no mode of action possessing the least plausibility had ever been devised, as the source of the required electrical power. *

(380.) It is not our intention, however, to enter into any further consideration of physiological subjects. They form, it is true, a most important and deeply interesting province of philosophical enquiry; but the view that we have taken of physical science has rather been directed to the study of inanimate nature, than to that of the mysterious phenomena of organization and life, which constitute the object of physiology. The history of the animal and vegetable productions of the globe, as affording objects and materials for the convenience and use of man, and as dependent on and indicative of the general laws which determine the distribution of heat, moisture, and other natural agents, over its surface, and the revolutions it has undergone, are of course intimately connected with our subject, and will, therefore, naturally afford room for some re-

* If the brain be an electric pile, constantly in action, it may be conceived to discharge itself at regular intervals, when the tension of the electricity developed reaches a certain point, along the nerves which communicate with the heart, and thus to excite the pulsations of that organ. This idea is forcibly suggested by a view of that elegant apparatus, the dry pile of Deluc; in which the successive accumulations of electricity are carried off by a suspended ball, which is kept by the discharges in a state of regular pulsation for any length of time. We have witnessed the action of such a pile maintained in this way for whole years in the study of the above-named eminent philosopher. The same idea of the cause of the pulsation of the heart appears to have occurred to Dr. Arnott; and is mentioned in his useful and excellent work on physics, to which however, we are not indebted for the suggestion, it having occurred to us independently many years ago.

marks, but not such as will long detain the reader's attention.

(381.) In *zoology*, the connection of peculiar modes of life and food, with peculiarities of structure, has given rise to systems of classification at once obvious and natural; and the great progress which has been made in comparative anatomy has enabled us to trace a graduated scale of organization almost through the whole chain of animal being; a scale not without its intervals, but which every successive discovery of animals heretofore unknown has tended to fill up. The wonders disclosed by microscopic observation have opened to us a new world, in which we discover, with astonishment, the extremes of minuteness and complexity of structure united; while, on the other hand, the examination of the fossil remains of a former state of creation has demonstrated the existence of animals far surpassing in magnitude those now living, and brought to light many forms of being which have nothing analogous to them at present, and many others which afford important connecting links between existing genera. And, on the other hand, the researches of the comparative anatomist and conchologist have thrown the greatest light on the studies of the geologist, and enabled him to discern, through the obscure medium of a few relics, scattered here and there through a stratum, circumstances connected with the formation of the stratum itself which he could have recognised by no other indication. This is one among many striking instances of the unexpected lights which sciences, however apparently remote, may throw upon each other.

(382.) To *botany* many of the same remarks apply. Its artificial systems of classification, however convenient, have not prevented botanists from endeavouring to group together the objects of their science in natural classes having a community of character more intimate than those which determine their place in the Linnean or any similar system; a community of character extending over the whole habit and properties of the individuals compared. The important chemical discoveries which have been lately made of peculiar proximate principles which, in an especial manner, characterize certain families of plants, hold out the prospect of a greatly increased field of interesting knowledge in this direction, and not only interesting, but in a high degree important, when it is considered that the principles thus brought into view are, for the most part, very powerful medicines, and are, in fact, the essential ingredients on which the medical virtues of the plants depend. The law of the distribution of the generic forms of plants over the globe, too, has, within a comparatively recent period, become an object of study to the naturalist; and its connection with the laws of climate constitutes one of the most interesting and important branches of natural-historical enquiry, and one on which great light remains to be thrown by future researches. It is this which constitutes the chief connecting link between botany and geology, and renders a knowledge of the vegetable fossils, of any portion of the earth's surface, indispensable to the formation of a correct judgment of the circumstances under which it existed in its ancient state. Fossil botany is accordingly cultivated with great and increasing ardour; and the subterraneous "Flora" of a geological formation is, in many instances, studied with a degree of care and precision little inferior to that which its surface exhibits.

CHAP. VI.

OF THE CAUSES OF THE ACTUAL RAPID ADVANCE OF THE PHYSICAL SCIENCES COMPARED WITH THEIR PROGRESS AT AN EARLIER PERIOD.

(383.) There is no more extraordinary contrast than that presented by the slow progress of the physical sciences, from the earliest ages of the world to the close of the sixteenth century, and the rapid developement they have since experienced. In the former period of their history, we find only small additions to the stock of knowledge, made at long intervals of time; during which a total indifference on the part of the mass of mankind to the study of nature operated to effect an almost complete oblivion of former discoveries, or, at best, permitted them to linger on record, rather as literary curiosities, than as possessing, in themselves, any intrinsic interest and importance. A few enquiring individuals, from age to age, might perceive their value, and might feel that irrepressible thirst after knowledge which, in minds of the highest order, supplies the absence both of external stimulus and opportunity. But the total want of a right direction given to enquiry, and of a clear perception of the objects to be aimed at, and the advantages to be gained by systematic and connected research, together with the general apathy of society to speculations remote from the ordinary affairs of life, and studiously kept involved in learned mystery, effectually prevented these occasional impulses from overcoming the inertia of ignorance, and impressing any regular and steady progress on science. Its objects, indeed, were confined in a region too sublime for vulgar comprehension. An earthquake, a comet, or a fiery meteor, would now and then call the attention of the whole world, and produce from all quarters a plentiful supply of crude and fanciful conjectures on their causes; but it was never supposed that sciences could exist among common objects, have a place among mechanical arts, or find worthy matter of speculation in the mine or the laboratory. Yet it cannot be supposed, that all the indications of nature continually passed unremarked, or that much good observation and shrewd reasoning on it failed to perish unrecorded, before the invention of printing enabled every one to make his ideas known to all the world. The moment this took place, however, the sparks of information from time to time struck out, instead of glimmering for a moment, and dying away in oblivion, began to accumulate into a genial glow, and the flame was at length kindled which was speedily to acquire the strength and rapid spread of a conflagration. The universal excitement in the minds of men throughout Europe, which the first out-break of modern science produced, has been already spoken of. But even the most sanguine anticipators could scarcely have looked forward to that steady, unintermitted progress which it has since maintained, nor to that rapid succession of great discoveries which has kept up the interest of the first impulse still vigorous and undiminished. It may truly, indeed, be said, that there is scarcely a single branch of physical enquiry which is either stationary, or which has not been, for many years past, in a constant state of advance, and in which the progress is not, at this moment, going on with accelerated rapidity.

(384.) Among the causes of this happy and desirable state of things, no doubt we are to look, in the first instance, to that great increase in wealth and civilization which has at once afforded the necessary leisure and diffused the taste for intellectual pursuits among numbers of mankind, which have long been and still continue steadily progressive in every principal European state, and which the increase and fresh establishment of civilized communities in every distant region are rapidly spreading over the whole globe. It is not, however, merely the increased number of cultivators of science, but their enlarged opportunities, that we have here to consider, which, in all those numerous departments of natural research that require local information, is in fact the most important consideration of all. To this cause we must trace the great extension which has of late years been conferred on every branch of natural history, and the immense contributions which have been made, and are daily making, to the departments of zoology and botany, in all their ramifications. It is obvious, too, that all the information that can possibly be procured, and reported, by the most enlightened and active travellers, must fall infinitely short of what is to be obtained by individuals actually resident upon the spot. Travellers, indeed, may make collections, may snatch a few hasty observations, may note, for instance, the distribution of geological formations in a few detached points, and now and then witness remarkable local phenomena; but the resident alone can make continued series of regular observations, such as the scientific determination of climates, tides, magnetic variations, and innumerable other objects of that kind, requires; can alone mark all the details of geological structure, and refer each stratum, by a careful and long continued observation of its fossil contents, to its true epoch; can alone note the habits of the animals of his country, and the limits of its vegetation, or obtain a satisfactory knowledge of its mineral contents, with a thousand other particulars essential to that complete acquaintance with our globe as a whole, which is beginning to be understood by the extensive designation of physical geography. Besides which, ought not to be omitted multiplied opportunities of observing and recording those extraordinary phenomena of nature which offer an intense interest, from the rarity of their occurrence as well as the instruction they are calculated to afford. To what, then, may we not look forward, when a spirit of scientific enquiry shall have spread through those vast regions in which the process of civilization, its sure precursor, is actually commenced and in active progress? And what may we not expect from the exertions of powerful minds called into action under circumstances totally different from any which have yet existed in the world, and over an extent of territory far surpassing that which has hitherto produced the whole harvest of human intellect? In proportion as the number of those who are engaged on each department of physical enquiry increases, and the geographical extent over which they are spread is enlarged, a proportionately increased facility of communication and interchange of knowledge becomes essential to the prosecution of their researches with full advantage. Not only is this desirable, to prevent a number of individuals from making the same discoveries at the same moment, which (besides the waste of valuable time) has always been a fertile source of jealousies and misunderstandings, by which great evils have been entailed on science; but because methods of observation are continually undergoing new improvements, or acquiring new facilities, a knowledge of which, it is for the general interest of science, should be diffused as widely and as rapidly as possible. By this means, too, a sense of common interest, of mutual assistance, and a feeling of sympathy in a common pursuit, are generated, which proves a powerful stimulus to exertion; and, on the other hand, means are thereby afforded of detecting and pointing out mistakes before it is too late for their rectification.

(385.) Perhaps it may be truly remarked, that, next to the establishment of institutions having either the promotion of science in general, or, what is still more practically efficacious in its present advanced state, that of particular departments of physical enquiry, for their express objects, nothing has exercised so powerful an influence on the progress of modern science as the publication of monthly and quarterly scientific journals, of which there is now scarcely a nation in Europe which does not produce several. The quick and universal circulation of these, places observers of all countries on the same level of perfect intimacy with each other's objects and methods, while the abstracts they from time to time (if well conducted) contain of the most important researches of the day consigned to the more ponderous tomes of academical collections, serve to direct the course of general observation, as well as to hold out, in the most conspicuous manner, models for emulative imitation. In looking forward to what may hereafter be expected from this cause of improvement, we are not to forget the powerful effect which must in future be produced by the spread of elementary works and digests of what is actually known in each particular branch of science. Nothing can be more discouraging to one engaged in active research, than the impression that all he is doing may, very likely, be labour taken in vain , that it may, perhaps, have been already done, and much better done, than, with his opportunities, or his resources, he can hope to perform it; and, on the other hand, nothing can be more exciting than the contrary impression. Thus, by giving a connected view of what has been done, and what remains to be accomplished in every branch, those digests and bodies of science, which from time to time appear, have, in fact, a very important weight in determining its future progress, quite independent of the quantity of information they communicate. With respect to elementary treatises it is needless to point out their utility, or to dwell on the influence which their actual abundance, contrasted with their past remarkable deficiency, is likely to exercise over the future. It is only by condensing, simplifying, and arranging, in the most lucid possible manner, the acquired knowledge of past generations, that those to come can be enabled to avail themselves to the full of the advanced point from which they will start.

(386.) One of the means by which an advanced state of physical science contributes greatly to accelerate and secure its further progress, is the exact knowledge acquired of physical data, or those normal quantities which we have more than once spoken of in the preceding pages (222.); a knowledge which enables us not only to appretiate the accuracy of experiments, but even to correct their results. As there is no surer criterion of the state of science in any age than the degree of care bestowed, and discernment exhibited, in the choice of such data, so as to afford the simplest possible grounds for the application of theories, and the degree of accuracy attained in their determination, so there is scarcely any thing by which science can be more truly benefited than by researches directed expressly to this object, and to the construction of tables exhibiting the true numerical relations of the elements of theories, and the actual state of nature, in all its different branches. It is only by such determinations that we can ascertain what changes are slowly and imperceptibly taking place in the existing order of things; and the more accurate they are, the *sooner* will this knowledge be acquired. What might we not now have known of the motions of the (so-called) fixed stars, had the ancients possessed the means of observation we now possess, and employed them as we employ them now?

(387.) In any enumeration of causes which have contributed to the recent rapid advancement of science, we must not forget the very important one of improved and constantly improving means of observation, both in instruments adapted for the exact measurement of quantity, and in the general convenience and well-judged adaptation to its purposes, of every description of scientific apparatus. In the actual state of science there are few observations which can be productive of any great advantage but such as afford accurate measurement; and an increased refinement in this respect is constantly called for. The degree of delicacy actually attained, we will not say in the most elaborate works of the highest art, but in such ordinary apparatus as every observer may now command, is such as could not have been arrived at unless in a state of the mechanical arts, which in its turn (such is the mutual re-action of cause and effect) requires for its existence a very advanced state of science. What an important influence may be exercised over the progress of a single branch of science by the invention of a ready and convenient mode of executing a definite measurement, and the construction and common introduction of an instrument adapted for it cannot be better exemplified than by the instance of the reflecting goniometer. This simple, cheap, and portable little instrument, has changed the face of mineralogy, and given it all the characters of one of the exact sciences.

(388.) Our means of perceiving and measuring minute quantities, in the important relations of weight, space, and time, seem already to have been carried to a point which it is hardly conceivable they should surpass. Balances have been constructed which have rendered sensible the millionth part of the whole quantity weighed; and to turn with the thousandth part of a grain is the performance of balances pretending to no very extraordinary degree of merit. The elegant invention of the sphærometer, by substituting the sense of touch for that of sight in the measurement of minute objects, permits the determination of their dimensions with a degree of precision which is fully adequate to the nicest purposes of scientific enquiry. By its aid an inch may be readily subdivided into ten or even twenty thousand parts; and the lever of contact, an instrument in use among the German opticians, enables us to appreciate quantities of space even yet smaller. For the subdivision of time, too, the perfection of modern mechanism has furnished resources which leave very little to be desired. By the aid of clocks and chronometers, as they are now constructed, a few tenths of a second is all the error that need be apprehended in the subdivision of a day; and for the further subdivision of smaller portions of time, instruments have been imagined which admit of almost unlimited precision, and permit us to appreciate intervals to the nicety of the hundredth, or even the thousandth part of a single second.* When the precision attainable by such means is

* See a description of a contrivance of this kind by Dr. Young, Lectures, vol. i. p. 191.

contrasted with what could be procured a few generations ago, by the rude and clumsy workmanship of even the early part of the last century, it will be no matter of astonishment that the sciences which depend on exact measurements should have made a proportional progress. Nor will any degree of nicety in physical determinations appear beyond our reach, if we consider the inexhaustible resources which science itself furnishes, in rendering the quantities actually to be determined by measure great multiples of the elements required for the purposes of theory, so as to diminish in the same proportion the influence of any errors which may be committed on the final results.

(389.) Great, indeed, as have been of late the improvements in the construction of instruments, both as to what regards convenience and accuracy, it is to the discovery of improved *methods* of observation that the chief progress of those parts of science which depend on exact determinations is owing. The balance of torsion, the ingenious invention of Cavendish and Coulomb, may be cited as an example of what we mean. By its aid we are enabled not merely to render sensible, but to subject to precise measurement and subdivision, degrees of force infinitely too feeble to affect the nicest balance of the usual construction, even were it possible to bring them to act on it. The galvanometer, too, affords another example of the same kind, in an instrument whose range of utility lies among electric forces which we have no other means of rendering sensible, much less of estimating with exactness. In determinations of quantities less minute in themselves, the methods devised by Messrs. Arago and Fresnel, for the measurement of the refractive powers of transparent media by means of the phenomenon of diffraction, may be cited as affording a degree of precision limited only by the wishes of the observer, and the time and patience he is willing to devote to his observation. And in respect of the direction of observations to points from which real information is to be obtained, and positive conclusions drawn, the hygrometer of Daniell may be cited as an elegant example of the introduction into general use of an instrument substituting an indication founded on strict principles for one perfectly arbitrary.

(390.) In speculating on the future prospects of physical science, we should not be justified in leaving out of consideration the probability, or rather certainty, of the occasional occurrence of those happy accidents which have had so powerful an influence on the past; occasions, where a fortunate combination opportunely noticed may admit us in an instant to the knowledge of principles of which no suspicion might occur but for some such casual notice. Boyle has entitled one of his essays thus remarkably,—" *Of Man's great Ignorance of the Uses of natural Things; or that there is no one Thing in Nature whereof the Uses to human Life are yet thoroughly understood.* * The whole history of the arts since Boyle's time has been one continued comment on this text; and if we regard among the uses of the works of nature, *that*, assuredly the noblest of all, which leads us to a knowledge of the Author of nature through the contemplation of the wonderful

* Boyle's Works, folio, vol. iii. Essay x. p. 185.

means by which he has wrought out his purposes in his works, the sciences have not been behind hand in affording their testimony to its truth. Nor are we to suppose that the field is in the slightest degree narrowed, or the chances in favour of such fortunate discoveries at all decreased, by those which have already taken place: on the contrary, they have been incalculably extended. It is true that the ordinary phenomena which pass before our eyes have been minutely examined, and those more striking and obvious principles which occur to superficial observation have been noticed and embodied in our systems of science; but, not to mention that by far the greater part of natural phenomena remain yet unexplained, every new discovery in science brings into view whole classes of facts which would never otherwise have fallen under our notice at all, and establishes relations which afford to the philosophic mind a constantly extending field of speculation, in ranging over which it is next to impossible that he should not encounter new and unexpected principles. How infinitely greater, for instance, are the mere chances of discovery in chemistry among the innumerable combinations with which the modern chemist is familiar, than at a period when two or three imaginary elements, and some ten or twenty substances, whose properties were known with an approach to distinctness, formed the narrow circle within which his ideas had to revolve? How many are the instances where a new substance, or a new property, introduced into familiar use, by being thus brought into relation with all our actual elements of knowledge, has become the means of developing properties and principles among the most common objects, which could never have otherwise been discovered? Had not platina (to take an instance) been an object of the most ordinary occurrence in a laboratory, would a suspicion have ever occurred that a lamp could be constructed to burn without flame; and should we have ever arrived at a knowledge of those curious phenomena and products of semi-combustion which this beautiful experiment discloses?

(391.) Finally, when we look back on what has been accomplished in science, and compare it with what remains to be done, it is hardly possible to avoid being strongly impressed with the idea that we have been and are still executing the labour by which succeeding generations are to profit.* In a few instances only have we arrived at those general axiomatic laws which admit of direct deductive inference, and place the solutions of physical phenomena before us as so many problems, whose principles of solution we fully possess, and which require nothing but acuteness of reasoning to pursue even into their farthest recesses. In fewer still have we reached that command of abstract reasoning itself which is necessary for the accomplishment of so arduous a task. Science, therefore, in relation to our faculties, still remains boundless and unexplored, and, after the lapse of a century and a half from the æra of Newton's discoveries, during which every department of it has been cultivated with a zeal and energy which have assuredly met their full return,

* Jackson, The Four Ages, p. 52. London: Cadell and Davies, 1798. 8vo.

we remain in the situation in which he figured himself,—standing on the shore of a wide ocean, from whose beach we may have culled some of those innumerable beautiful productions it casts up with lavish prodigality, but whose acquisition can be regarded as no diminution of the treasures that remain.

(392.) But this consideration, so far from repressing our efforts, or rendering us hopeless of attaining any thing intrinsically great, ought rather to excite us to fresh enterprise, by the prospect of assured and ample recompense from that inexhaustible store which only awaits our continued endeavours. "It is no detraction from human capacity to suppose it incapable of infinite exertion, or of exhausting an infinite subject."* In whatever state of knowledge we may conceive man to be placed, his progress towards a yet higher state need never fear a check, but must continue till the last existence of society.

(393.) It is in this respect an advantageous view of science, which refers all its advances to the discovery of general laws, and to the inclusion of what is already known in generalizations of still higher orders; inasmuch as this view of the subject represents it, as it really is, essentially incomplete, and incapable of being fully embodied in any system, or embraced by any single mind. Yet it must be recollected that, so far as our experience has hitherto gone, every advance towards generality has at the same time been a step towards simplification. It is only when we are wandering and lost in the mazes of particulars, or entangled in fruitless attempts to work our way downwards in the thorny paths of

* Jackson, The Four Ages, p. 90.

applications, to which our reasoning powers are incompetent, that nature appears complicated: — the moment we contemplate it as it is, and attain a position from which we can take a commanding view, though but of a small part of its plan, we never fail to recognise that sublime simplicity on which the mind rests satisfied that it has attained the truth.

INDEX.

THE END.